ISNM
International Series of Numerical Mathematics
Vol. 137

Recent Progress in Multivariate Approximation

4th International Conference, Witten-Bommerholz (Germany), September 2000

Werner Haussmann
Kurt Jetter
Manfred Reimer
Editors

Springer Basel AG

Editors:

Werner Haussmann
Mathematisches Institut
Gerhard-Mercator-Universität Duisburg
47048 Duisburg
Germany
haussmann@math.uni-duisburg.de

Kurt Jetter
Institut für Angewandte Mathematik und Statistik
Universität Hohenheim
70593 Stuttgart
Germany
jetter@uni-hohenheim.de

Manfred Reimer
Fachbereich Mathematik
Universität Dortmund
44221 Dortmund
Germany
reimer@euler.mathematik.uni-dortmund.de

2000 Mathematics Subject Classification 41-06; 41A63, 31B05, 68U10

A CIP catalogue record for this book is available from the Library of Congress, Washington D.C., USA

Deutsche Bibliothek Cataloging-in-Publication Data

Recent progress in multivariate approximation : 4th international
conference, Witten-Bommerholz (Germany), September 2000 / Werner Haussmann
... ed.. - Basel ; Boston ; Berlin : Birkhäuser, 2001
   (International series of numerical mathematics ; Vol. 137)
   ISBN 978-3-0348-9498-2   ISBN 978-3-0348-8272-9 (eBook)
   DOI 10.1007/978-3-0348-8272-9

© 2001 Springer Basel AG
Originally published by Birkhäuser Verlag in 2001
Softcover reprint of the hardcover 1st edition 2001

Printed on acid-free paper produced of chlorine-free pulp. TCF ∞
ISBN 978-3-0348-9498-2

9 8 7 6 5 4 3 2 1

Preface

These proceedings contain the main topics and results presented at the Fourth International Conference on Multivariate Approximation. The meeting took place during the week of September 24–29, 2000 at the now well–known "Haus Bommerholz", the guest–house of the University of Dortmund. It hosted 43 participants from 16 countries, and the program included 9 invited one–hour lectures, 21 contributed talks, and two problem sessions. The articles collected here are carefully peer–refereed and suitably edited for publication.

Following the tradition of this series of conferences, the meeting was aimed at advancing selected topics of Multivariate Approximation Theory. These include approximation on compact sets (such as spheres, balls, or compact homogeneous manifolds), spherical designs and energy functionals, interpolation by radial basis functions and by splines, frame theory and Gabor analysis, refinable function systems and subdivision, properties of harmonic, polyharmonic and blending functions, sampling and data compression, among others.

The editors would like to express their thanks to all who have given their support to the conference, and to the preparation of this book. In particular, our thanks go to the Deutsche Forschungsgemeinschaft for their funding of the conference, to the University of Dortmund, and to the staff of Haus Bommerholz for their help running the conference, and to the Universities of Dortmund and Duisburg for the financial support of this proceedings volume. We are grateful to all the authors and referees who helped us maintain a high scientific standard. In addition, we acknowledge the constant help of Dr. Hermann Hoch (Duisburg) during the production of this proceedings volume. Last but not least, we would like to thank the publisher for including this book in the *International Series of Numerical Mathematics*.

June 30, 2000

Werner Haußmann

Kurt Jetter

Manfred Reimer

Acknowledgment

We are deeply indebted to the

Deutsche Forschungsgemeinschaft (DFG)

whose generous funding made this conference possible.

The production of this volume was jointly sponsored by the

Universität Dortmund

and by the

Gerhard–Mercator–Universität Duisburg.

The Editors

Table of Contents

8   Table of Contents

Participants

Nicolay N. Andreev, *Department of Function Theory, Steklov Institute of Mathematics, stz. Gubkina 8, Moscow GSP–1, Russia*
email: andreev@mccme.ru

David H. Armitage, *Department of Pure Mathematics, Queen's University, Belfast BT7 1NN, Northern Ireland*
email: d.armitage@qub.ac.uk

John J. Benedetto, *Department of Mathematics, University of Maryland, College Park, MD 20742, U. S. A.*
email: jjb@math.umd.edu

Elena E. Berdysheva, *Mathematisches Institut, Universität Erlangen, Bismarckstr. 1 1/2, D–91054 Erlangen, Germany*
email: berdyshe@mi.uni-erlangen.de

Peter Binev, *Institute of Mathematics, Bulgarian Academy of Sciences, Sofia 1090, Bulgaria*
email: binev@fmi.uni-sofia.bg

Borislav D. Bojanov, *Department of Mathematics, University of Sofia, 5 James Bourchier Blvd., BG–1164 Sofia, Bulgaria*
email: boris@fmi.uni-sofia.bg

Len P. Bos, *Department of Mathematics and Statistics, University of Calgary, Calgary, Alberta T2N 1N4, Canada*
email: lpbos@math.ucalgary.ca

Dietrich Braess, *Institut für Mathematik, Ruhr-Universität Bochum,*
  D–44780 Bochum, Germany
  email: braess@num.ruhr-uni-bochum.de

Johann Brauchart, *Institut für Mathematik A, Technische Universität Graz,*
  Steyrergasse 30, A-8010 Graz, Austria
  email: brauchart@finanz.math.tu-graz.ac.at

Andreas Busch, *Fachbereich Mathematik, Universität Dortmund,*
  D–44221 Dortmund, Germany

Maria Charina, *Institut für Angewandte Mathematik und Statistik,*
  Universität Hohenheim, D–70593 Stuttgart, Germany
  email: mcharina@uni-hohenheim.de

Ole Christensen, *Department of Mathematics, Technical University of*
  Denmark, DK–2800 Lyngby, Denmark
  email: olechr@mat.dtu.dk

Costanza Conti, *Dipartimento di Energetica, Università di Firenze,*
  Via Lombroso 6 / 17, I–2800 Firenze, Italy
  email: costanza@sirio.de.unifi.it

Franz–Jürgen Delvos, *Fachbereich Mathematik, Universität Siegen,*
  Walter–Flex–Str. 33, D–57068 Siegen, Germany
  email: delvos@mathematik.uni-siegen.de

Zeev Ditzian, *Department of Mathematics, University of Alberta,*
  Edmonton, Alberta, Canada T6G 2G1
  email: zditzian@math.ualberta.ca

Nira Dyn, *School of Mathematical Sciences, Tel Aviv University,*
Ramat Aviv, 69978 Tel Aviv, Israel
email: niradyn@math.tau.ac.il

Karol Dziedziul, *Faculty of Applied Mathematics, Technical University*
of Gdansk, ul. Narutowicza 11/12, 80-952 Gdansk, Poland
email: kdz@mifgate.pg.gda.pl

Hans G. Feichtinger, *Institut für Mathematik, Universität Wien,*
Strudlhofgasse 4, A–1090 Wien, Austria
email: fei@tyche.mat.univie.ac.at

Jörg Fliege, *Fachbereich Mathematik, Universität Dortmund,*
D–44221 Dortmund, Germany
email: fliege@math.uni-dortmund.de

Stephen J. Gardiner, *Department of Mathematics,*
University College Dublin, Dublin 4, Ireland
email: gardiner@acadamh.ucd.ie

Peter J. Grabner, *Institut für Mathematik A,*
Technische Universität Graz, Steyrergasse 30, A–8010 Graz, Austria
email: grabner@weyl.math.tu-graz.ac.at

Werner Haußmann, *Mathematisches Institut, Gerhard–Mercator–Universität*
Duisburg, D–47048 Duisburg, Germany
email: haussmann@math.uni-duisburg.de

Simon Hubbert, *Department of Mathematics, Huxley Building,*
Imperial College, 180 Queen's Gate, London SW7 2BZ, England
email: simon.hubbert@ic.ac.uk

Kurt Jetter, *Institut für Angewandte Mathematik und Statistik,*
   Universität Hohenheim, D–70593 Stuttgart, Germany
   email: kjetter@uni-hohenheim.de

Anna Kamont, *Mathematical Institute PAN, ul. Abrahama 18,*
   81–825 Sopot, Poland
   email: a.kamont@impan.gda.pl

John Klinkhammer, *Mathematisches Institut, Gerhard–Mercator–Universität*
   Duisburg, D–47048 Duisburg, Germany
   email: klinkhammer@math.uni-duisburg.de

Hermann König, *Mathematisches Seminar, Universität Kiel,*
   Ludewig–Meyn–Str. 4, D–24098 Kiel, Germany
   email: hkoenig@math.uni-kiel.de

Alain Le Méhauté, *Département de Mathématiques, Université de Nantes,*
   2 Rue de la Houssinière, F–44072 Nantes Cedex, France
   email: alm@math.univ-nantes.fr

Jeremy Levesley, *Department of Mathematics, University of Leicester,*
   Leicester LE1 7RH, England
   email: jl1@mcs.le.ac.uk

Tom Lyche, *Institutt for Informatikk, University of Oslo,*
   P.O.Box 1080 Blindern, N–0316 Oslo, Norway
   email: tom@ifi.uio.no

Ulrike Maier, *Mathematisches Institut, Justus–Liebig–Universität Giessen,*
   Arndtstr. 2, D–35392–Giessen, Germany
   email: ulrike.maier@math.uni-giessen.de

H. Michael Möller, *Fachbereich Mathematik, Universität Dortmund,*
D–44221 Dortmund, Germany
email: moeller@math.uni-dortmund.de

Geno Nikolov, *Department of Mathematics, University of Sofia,*
5 James Bourchier Blvd., BG–1164 Sofia, Bulgaria
email: geno@fmi.uni-sofia.bg

Günther Nürnberger, *Fakultät für Mathematik und Informatik,*
Universität Mannheim, D–68131 Mannheim, Germany
email: nuern@euklid.math.uni-mannheim.de

Petar Petrov, *Department of Mathematics, University of Sofia,*
5 James Bourchier Blvd., BG–1164 Sofia, Bulgaria
email: peynov@fmi.uni-sofia.bg

Götz Pfander, *Institut für Angewandte Mathematik und Statistik,*
Universität Hohenheim, D–70593 Stuttgart, Germany
email: pfander@uni-hohenheim.de

Manfred Reimer, *Fachbereich Mathematik, Universität Dortmund,*
D–44221 Dortmund, Germany
email: reimer@math.uni-dortmund.de

Edward B. Saff, *Department of Mathematics,*
University of South Florida, Tampa FL 33620, U.S.A.
email: esaff@math.usf.edu

Robert Schaback, *Institut für Numerische und Angewandte Mathematik,*
Universität Göttingen, D–37083 Göttingen, Germany
email: schaback@math.uni-goettingen.de

Robert S. Womersley, *School of Mathematics, University of New South Wales, Sydney, NSW 2052, Australia*
email: r.womersley@unsw.edu.au

Vladimir A. Yudin, *Department of Higher Mathematics, Moscow Institute of Power Engineering, Krasnokazarmennaya 15, Moscow 105 835, Russia*
email: yudin@vayudin.mccme.ru

Frank Zeilfelder, *Fakultät für Mathematik und Informatik, Universität Mannheim, D–68131 Mannheim, Germany*
email: zeilfeld@euklid.math.uni-mannheim.de

Georg Zimmermann, *Institut für Angewandte Mathematik und Statistik, Universität Hohenheim, D–70593 Stuttgart, Germany*
email: gzim@uni-hohenheim.de

Scientific Program of the Conference

Monday, September 25:

1st Morning Session: Chair: *K. Jetter*

9:00	Opening of the Conference
9:30 – 10:30	*J. J. Benedetto*: The Role of Tiling in Multidimensional Fourier Frames and Wavelet Theory

2nd Morning Session: Chair: *H. König*

11:00 – 12:00	*N. Dyn*: Multiscale Subdivision Schemes and Refinability
12:00 – 12:30	*C. Conti*: Convolution of Function Vectors

Afternoon Session: Chair: *L. Bos*

15:30 – 16:30	*J. Levesley*: Approximation on Compact Homogeneous Manifolds
16:30 – 17:00	*S. Hubbert*: Interpolation on Spheres Using Basis Function Methods
17:00 – 17:30	*A. Le Méhauté* : On the Evaluation of Splines and Radial Basis Functions
17:30 – 18:00	*F. Zeilfelder*: Local Lagrange Interpolation by Bivariate Splines
19:00 – 20:30	1st Problem Session

Tuesday, September 26:

1st Morning Session: Chair: *G. Nürnberger*

9:00 – 10:00   *E. Saff*: Distributing Many Points on a Sphere:
                 A Minimum Energy Approach

10:00 – 10:30   *P. J. Grabner*: Energy Functionals and
                 Numerical Integration

2nd Morning Session: Chair: *P. Binev*

11:00 – 11:30   *N. A. Andreev*: Approximations on the
                 m−Dimensional Sphere

11:30 – 12:00   *V. A. Yudin*: New Properties of Spherical Designs

12:00 – 12:30   *U. Maier*: Interpolation at Spherical Designs

Afternoon Session: Chair: *N. Dyn*

15:30 – 16.30   *Z. Ditzian*: New Moduli of Smoothness on the Sphere

16:30 – 17:00   *E. E. Berdysheva*: Several Extremal Problems
                 for Multivariate Positive Definite Functions

17:00 – 17:30   *T. Lyche*: On the p−Norm Condition Number
                 of the Multivariate Triangular Bernstein Basis

Wednesday, September 27:

1st Morning Session: Chair: *E. Saff*

9:00 – 9:30	*P. Petrov*: Best One–Sided L^1–Approximation by $B^{2,1}$–Blending Functions
9:30 – 10:00	*D. H. Armitage*: Harmonic Functions that Vanish at Certain Lattice Points
10:00 – 10:30	*S. J. Gardiner*: Convergence of Rational Interpolants with Preassigned Poles

2nd Morning Session: Chair: *M. Reimer*

11:30 – 12:30	*R. S. Womersley*: Choosing Points on the Sphere for Polynomial Interpolation and Quadrature

Thursday, September 28:

1st Morning Session: Chair: *A. Le Méhauté*

9:00 – 10:00	*B. D. Bojanov*: Cubature Formulae for Polyharmonic Functions
10:00 – 10:30	*G. Nikolov*: Some Cubature Formulae Using Mixed Type Data

2nd Morning Session: Chair: *T. Lyche*

11:00 – 11:30	*A. Kamont*: General Haar Systems and Greedy Approximation
11:30 – 12:00	*P. G. Binev*: Tree Approximation
12:00 – 12:30	*G. Pfander*: Multidimensional Generalized Haar Wavelets

Afternoon Session: Chair: *J. J. Benedetto*

15:30 – 16:30   *H. G. Feichtinger*: Functional Analytic Concepts for
                   Spline Type Spaces and Gabor Analysis

16:30 – 17:00   *G. Zimmermann*: Short–Time Fourier Transform
                   of Ultra–Distributions

17:00 – 17:30   *O. Christensen*: Approximation of the Inverse Frame
                   Operator and Applications to Gabor Frames

19:00 – 20:30   2nd Problem Session

Friday, September 29:

1st Morning Session: Chair: *H. M. Möller*

 9:30 – 10:00   *K. Dziedziul*: Asymptotic Formulas for the Error
                   in Cardinal Interpolation and Orthogonal Projection

10:00 – 10:30   *M. Charina*: Image Data Compression with
                   Three–Directional Splines

2nd Morning Session: Chair: *W. Haußmann*

11:00 – 12:00   *L. Bos*: Complex Variable Techniques in
                   Multivariate Approximation

12:00   Closing of the Conference

Jochen W. Schmidt in Memoriam

Jochen W. Schmidt
August 8, 1931 – May 14, 2000

Prof. Dr. Jochen W. Schmidt died on May 14, 2000. His death came unexpectedly while he was still actively serving the mathematical community just before leaving for the conference on approximation theory in Nashville, Tennessee.

Jochen W. Schmidt was born at Neukloster in Mecklenburg on August 8, 1931. He was married and leaves behind his spouse, four children and six granddaughters. He studied mathematics at the Universities of Rostock and Greifswald, where he passed his teacher's qualifications. In 1956 he joined Fr. A. Willers and H. Heinrich as an assistant at the Technische Hochschule Dresden. These able and dedicated teachers inspired him and encouraged him to apply functional analytic methods to his research in numerical mathematics. This approach was new at the time. At the invitation of L. Collatz he spent a research semester at the University of Hamburg which turned out to be his last opportunity to visit the West for thirty years.

In 1959 Jochen W. Schmidt obtained his doctoral degree, in 1964 he obtained his habilitation. In 1967 he was appointed professor at the Technische Universität Dresden, soon afterwards he was promoted to full professor of numerical mathematics. So Dresden with its intellectual flair became his home until his death.

Jochen W. Schmidt taught students of mathematics, science and engineering in applied mathematics, all this very successfully. He also conducted very extensive and serious research work. These activities earned him broad international recognition. As a visible sign of this he was rewarded the Medal of the University of Helsinki and was invited to join the editorial boards of *Zeitschrift für Angewandte Mathematik und Mechanik*, of *Computing*, and of *Numerische Mathematik*. His acceptance against the objections of GDR officials complicated his life considerably. In 1977 he became the head of the group on Numerical Mathematics. But his open and unequivocal stance brought him more and more into conflict with the political system. So he had to resign from this administrative position and even feared for his professorship as he became the target of Stasi informers.

But Jochen W. Schmidt increased his activities in teaching and research, focussing on new research fields such as optimisation and approximation theory and at the end of his life had supervised a total of 13 PhD theses.

After the Wall came down in 1989 he seized the long-awaited opportunity to travel in Western countries and to attend important mathematical conferences worldwide usually presenting talks. He stayed involved in various academic committees in Dresden and played a leading role in the renewal of the Department of Mathematics. His department organized an impressive scientific colloquium for his 65th birthday and published a *Festschrift* comprising a synopsis of his life and a list of his about 150 publications to honour his faithful contributions to the department.

Jochen W. Schmidt was twice selected by the mathematical community as a leading mathematics reviewer of the German science foundation (DFG). This great honour meant a lot of hard work for him during the last decade of his life.

Jochen W. Schmidt participated in the preceding Bommerholz Conferences on Multivariate Approximation and contributed to their success by presentations on scattered data and shape preserving interpolation by bivariate spline functions.

We are mourning a mathematician of high scientific merit and a trustworthy colleague and friend.

Manfred Reimer, Dortmund   Hubert Schwetlick, Dresden

As appendix to this proceedings volume the list of publications and further data on J. W. Schmidt are enclosed.

International Series of Numerical Mathematics
Vol. 137, ©2001 Birkhäuser Verlag Basel/Switzerland

Best Approximation of Polynomials on the Sphere and on the Ball

Nikolay N. Andreev and Vladimir A. Yudin[1]

Abstract

In this paper an approach for finding polynomials of minimum deviation from zero on the sphere and on the ball of n–dimensional Euclidean space is described.

Introduction

In this paper we study the problem of best approximation to functions of several variables by algebraic polynomials on the sphere and on the ball. A classical theorem is due to P. L. Chebyshev:

For every $q \in \mathbb{N}$ the following result for univariate polynomials holds true :

$$\inf_{c_k} \max_{-1 \leq x \leq 1} \left| x^q - \sum_{k=0}^{q-1} c_k x^k \right| = \frac{1}{2^{q-1}} \ .$$

In order to describe corresponding results in the multivariate setting, let

$$S^{m-1} = \{x \in \mathbb{R}^m : x_1^2 + \ldots + x_m^2 = 1\}, \qquad B^m = \{x \in \mathbb{R}^m : x_1^2 + \ldots + x_m^2 \leq 1\},$$

be the unit sphere and the unit ball of $\mathbb{R}^m$, respectively. By $\mathcal{P}_n^m$ we denote the set of algebraic polynomials in m variables $x = (x_1, \ldots, x_m)$ of degree less or equal to $n \in \mathbb{N}_0$:

$$\mathcal{P}_n^m = \left\{ \sum_{0 \leq \nu_1 + \ldots + \nu_m \leq n} c_\nu \, x_1^{\nu_1} \ldots x_m^{\nu_m} \right\}.$$

We shall use the notation $x^\nu = x_1^{\nu_1} \ldots x_m^{\nu_m}$, $|\nu| = \nu_1 + \ldots + \nu_m$. Given a continuous function f, $f(x) = f(x_1, \ldots, x_m)$, we denote by

$$E_{n-1}(f)_{C[S^{m-1}]} = \inf_{p \in \mathcal{P}_{n-1}^m} \|f(x) - p(x)\|_{C[S^{m-1}]}$$

---

[1]The work was partualy support by RFFI-GFEN N 99-01-39006

and

$$E_{n-1}(f)_{C[B^m]} = \inf_{p \in \mathcal{P}^m_{n-1}} \|f(x) - p(x)\|_{C[B^m]},$$

the minimum distance of f from $\mathcal{P}^m_{n-1}$ in the Chebyshev norm on the sphere and on the ball, respectively. We write $E_{n-1}(f)_C$ when we speak about both cases S^{m-1} and B^m.

This paper is devoted to the problem of finding $E_{n-1}(x_1^{\nu_1} \cdot \ldots \cdot x_m^{\nu_m})_C$. Results concerning approximations on the cube can be found in [6, 4]. Originally our interest was directed to the sphere only, but later we were stimulated to investigate approximations on the ball by the work of M. Reimer [4] and by his lecture at a former Bommerholz conference [5].

In the case of dimension $m = 2$ Gearhart [3] proved that the polynomials

$$\{U_{n_1}(x_1)\, U_{n_2}(x_2) + U_{n_1-2}(x_1)\, U_{n_2-2}(x_2)\}$$

$$= 2^{n_1+n_2} x_1^{n_1} x_2^{n_2} + \text{terms of lower degree}$$

are polynomials of minimum deviation from zero on $C[B^2]$ with the considered leading coefficient. Here U_n are the Chebyshev polynomials of the second kind with the convention that $U_{-1} := 0$, $U_{-2} := -1$. It follows immediately that

$$E_{|n|-1}(x_1^{n_1} x_2^{n_2}) = \frac{1}{2^{|n|-1}}.$$

1   The Case of Lower Dimension

It suffices to consider the case $n_1 \cdot \ldots \cdot n_m \neq 0$, only. For if one of the numbers $n_1, \ldots, n_m$ is equal to zero then the problem reduces to a problem in lower dimension. Indeed, let the function $f \in C[B^m]$ depend only on k ($k \leq m$) variables $x_1, \ldots, x_k$. As $B^k \subset B^m$, we get for any $p \in \mathcal{P}^m_{n-1}$

$$\|f(x) - p(x)\|_{C[B^m]} \quad \geq \quad \|f(x) - p(x_1, \ldots, x_k, 0, \ldots, 0)\|_{C[B^k]}.$$

It is obvious that

$$\mathcal{P}^k_{n-1} = \{p(x_1, \ldots, x_k, 0, \ldots, 0) \mid p \in \mathcal{P}^m_{n-1}\}.$$

Therefore we get for $k < m$

$$E_{n-1}(f)_{C[B^m]} = \inf_{p \in \mathcal{P}^m_{n-1}} \|f - p\|_{C[B^m]} \geq \inf_{p \in \mathcal{P}^k_{n-1}} \|f - p\|_{C[B^k]} = E_{n-1}(f)_{C[B^k]}.$$

From the results of Chebyshev and Gearhart we have

$$E_{n-1}(x_1^n)_{C[B^m]} \geq \frac{1}{2^{n-1}}, \qquad n \in \mathbb{N}, \; m \geq 1;$$

$$E_{n-1}(x_1^{n_1} x_2^{n_2})_{C[B^m]} \geq \frac{1}{2^{n-1}}, \qquad n_1, n_2 \in \mathbb{N}, \; m \geq 2.$$

Upper estimates

$$E_{n-1}(x_1^n)_{C[B^m]} \leq \frac{1}{2^{n-1}}, \qquad E_{n-1}(x_1^{n_1} x_2^{n_2})_{C[B^m]} \leq \frac{1}{2^{n-1}}$$

are provided by the choices

$$x_1^n - p(x) = \frac{1}{2^{n-1}} T_n(x_1);$$

and

$$x_1^{n_1} x_2^{n_2} - p(x) = \frac{1}{2^{n_1+n_2}} \{ U_{n_1}(x_1) U_{n_2}(x_2) + U_{n_1-2}(x_1) U_{n_2-2}(x_2) \},$$

where T_n, U_n are the Chebyshev polynomials of first and of second kind, respectively. Recall $U_{-1} = 0$, $U_{-2} = -1$.

2 Results

In this section we state our main results. The proofs of Theorems 2.1, 2.2 and 2.3 are given in [1], the proof of Theorem 2.4 in Section 3.

Theorem 2.1. *Let $m \in \mathbb{N}$, $m \geq 2$. Then*

$$E_{m-1}(x_1 \cdot \ldots \cdot x_m)_{C[B^m]} = E_{m-1}(x_1 \cdot \ldots \cdot x_m)_{C[S^{m-1}]} = \|x_1 \cdot \ldots \cdot x_m\|_{C[S^{m-1}]} = m^{-m/2}.$$

$$E_3(x_1^4 + \ldots + x_m^4)_{C[B^m]} = E_3(x_1^4 + \ldots + x_m^4)_{C[S^{m-1}]} = \left\| x_1^4 + \ldots + x_m^4 - \frac{m+1}{2m} \right\|_{C[S^{m-1}]}.$$

$$E_{m+3}((x_1^4 + \ldots + x_m^4)x_1 \ldots x_m)_{C[B^m]} = \inf_{c \in \mathbb{R}} \|(x_1^4 + \ldots + x_4^4 - c)x_1 \ldots x_m\|_{C[B^m]}.$$

$$E_{m+3}((x_1^4 + \ldots + x_m^4)x_1 \ldots x_m)_{C[S^{m-1}]} = \inf_{c \in \mathbb{R}} \|(x_1^4 + \ldots + x_4^4 - c)x_1 \ldots x_m\|_{C[S^{m-1}]}.$$

Theorem 2.2. *Let* $m = 3$, $\tau = (\sqrt{5}+1)/2$, $M = \{(1,2),(2,3),(3,1)\}$. *Then*

$$E_5\left(\sum_{(i,j)\in M} (\tau x_i + x_j)^6 + (\tau x_i - x_j)^6 \right)_{C[S^{m-1}]}$$

$$= \inf_{c\in\mathbb{R}} \left\| \sum_{(i,j)\in M} (\tau x_i + x_j)^6 + (\tau x_i - x_j)^6 - c \right\|_{C[S^{m-1}]}.$$

Theorem 2.3. *Let* $m = 4$. *Then*

$$E_5\left(\sum_{1\le i<j\le 4} ((x_i + x_j)^6 + (x_i - x_j)^6) \right)_{C[S^{m-1}]}$$

$$= \inf_{c\in\mathbb{R}} \left\| \sum_{1\le i<j\le 4} ((x_i + x_j)^6 + (x_i - x_j)^6) - c \right\|_{C[S^{m-1}]}.$$

For $m \ge 2$ we denote by $a = a(m)$ the solution of the equation

$$\frac{2}{m+1}\left(\frac{m-1}{m+1}\right)^{\frac{m-1}{2}} (1-a)^{\frac{m-1}{2}} = a \tag{2.1}$$

which belongs to the interval $(0,1)$. The first of these values are $a(2) = 1/4$, $a(3) = 3 - \sqrt{8}, \ldots$. For $m \to \infty$ we get from (2.1)

$$a(m) = \frac{\theta}{m}\{1 + o(1)\},$$

where θ is the solution of the transcendental equation $\theta e^{\theta/2} = 2/e$.

Theorem 2.4. *Let* $m \in \mathbb{N}$, $m \ge 2$. *Then*

$$E_m(x_1^2 x_2 x_3 \ldots x_m)_{C[B^m]} = E_m(x_1^2 x_2 x_3 \ldots x_m)_{C[S^{m-1}]}$$

$$= \|(x_1^2 - a(m))x_2 x_3 \ldots x_m\|_C = \frac{a(m)}{(m-1)^{\frac{m-1}{2}}}.$$

3 Proof of Theorem 2.4

Proof of Theorem 2.4. We put

$$F(x, \lambda) = x_1^2 x_2 x_3 \ldots x_m - \lambda \, x_2 x_3 \ldots x_m.$$

The first step is to show that

$$E_m(x_1^2 x_2 x_3 \ldots x_m)_{C[S^{m-1}]} = \inf_{\lambda \in \mathbb{R}} \|F(x, \lambda)\|_{C[S^{m-1}]}. \tag{3.1}$$

Indeed, as $x_1^2 x_2 \ldots x_m$ is even with respect to x_1 and odd with respect to $x_2, \ldots, x_m$, a polynomial $p^* \in \mathcal{P}_m^m$ of best uniform approximation exists which is also even in x_1 and odd in $x_2, \ldots, x_m$. So it must have the form $p^*(x) = \lambda \cdot x_2 \cdots x_m$ with some constant λ. This proves (3.1).

Now denote

$$I_m = \inf_{\lambda \in \mathbb{R}} \|F(x, \lambda)\|_{C[S^{m-1}]}, \qquad J_m = \inf_{\lambda \in \mathbb{R}} \|F(x, \lambda)\|_{C[B^m]}.$$

We shall estimate I_m from below and J_m from above. It is obvious that $I_m \leq J_m$. Our aim is to show that they are equal: $I_m = J_m$.

First we estimate J_m from above. Take $\lambda = a(m)$. For any $x \in B^m$

$$x_2^2 + \ldots + x_m^2 \leq 1 - x_1^2.$$

By the inequalities between the arithmetic mean and geometric mean we get

$$|x_2 \ldots x_m| = |x_2^2 \ldots x_m^2|^{1/2} \leq \left(\frac{x_1^2 + \ldots + x_m^2}{m-1} \right)^{\frac{m-1}{2}} \leq \left(\frac{1}{m-1} \right)^{\frac{m-1}{2}} (1 - x_1^2)^{\frac{m-1}{2}}.$$

That is why for any $x \in B^m$

$$J_m \leq \left(\frac{1}{m-1} \right)^{\frac{m-1}{2}} \max_{x \in B^m} \left| (x_1^2 - a)(1 - x_1^2)^{\frac{m-1}{2}} \right|. \tag{3.2}$$

Recall the abbreviation $a = a(m)$. The function $\Phi(t) = |t^2 - a| (1 - t^2)^{\frac{m-1}{2}}$ has 3 points of local maximum on the segment $[-1, 1]$, namely $0, \pm t^* = \sqrt{\frac{2 + (m-1)a}{m+1}}$. Since

$$\Phi(0) = a, \qquad \Phi(\pm t^*) = \frac{2}{m+1} \left(\frac{m-1}{m+1} \right)^{\frac{m-1}{2}} (1 - a)^{\frac{m+1}{2}},$$

it follows from (2.1) that

$$\Phi(0) = \Phi(\pm t^*) = a.$$

From (3.2) we have

$$J_m \leq \left(\frac{1}{m-1}\right)^{\frac{m-1}{2}} a.$$

In order to get a lower estimate of I_m let us take two points

$$M_1 = \left(0, \sqrt{\frac{1}{m-1}}, \dots, \sqrt{\frac{1}{m-1}}\right),$$

$$M_2 = \left(\sqrt{\frac{2+(m-1)a}{m+1}}, \sqrt{\frac{1-a}{m+1}}, \dots, \sqrt{\frac{1-a}{m+1}}\right).$$

As a consequence of (2.1) we have $M_2 \in S^{m-1}$, and it is obvious that $M_1 \in S^{m-1}$. For any $\lambda \in \mathbb{R}$ we get

$$\|F(x,\lambda)\|_{C[S^{m-1}]} \geq F(M_1, \lambda) = |\lambda| \left(\frac{1}{m-1}\right)^{\frac{m-1}{2}}$$

and

$$\|F(x,\lambda)\|_{C[S^{m-1}]} \geq F(M_2, \lambda) = \left|\frac{2+(m-1)a}{m+1} - \lambda\right| \left(\frac{1-a}{m+1}\right)^{\frac{m-1}{2}}.$$

Together this yields

$$(m-1)^{\frac{m-1}{2}} \|F(x,\lambda)\|_{C[S^{m-1}]} + \left(\frac{m+1}{1-a}\right)^{\frac{m-1}{2}} \|F(x,\lambda)\|_{C[S^{m-1}]} \geq$$

$$\geq |\lambda| + \left|\frac{2+(m-1)a}{m+1} - \lambda\right| \geq \frac{2+(m-1)a}{m+1},$$

and we obtain

$$\|F(x,\lambda)\|_{C[S^{m-1}]} \geq \frac{2+(m-1)a}{m+1} \left\{(m-1)^{\frac{m-1}{2}} + \left(\frac{m+1}{1-a}\right)^{\frac{m-1}{2}}\right\}^{-1}.$$

Using (2.1) one can calculate that the right hand of the inequality is equal to $a(\frac{1}{m-1})^{\frac{m-1}{2}}$ which is equal to the upper estimate of J_m. So we proved that

$$\|F(x,\lambda)\|_{C[S^{m-1}]} = \|F(x,\lambda)\|_{C[B^m]} = \left(\frac{1}{m-1}\right)^{\frac{m-1}{2}} a.$$

The statement of the theorem follows from (3.1).

Remark 1. Further theorems of this kind can be proved by the method described in [1]. Theorem 2.4 shows that classical Chebyshev polynomials as polynomials of minimum deviation from zero on the sphere and the ball of m–dimensional Euclidean space occur only in certain cases. In more general m–dimensional approximation problems non–classical polynomials will be extremal.

Remark 2. The problem treated above suggests the following question:

What is the minimum deviation,

$$\inf_{c_k} \max_{-1 \le x \le 1} |(x^n + c_{n-1}x^{n-1} + \ldots + c_0)w(x)|,$$

in the wheighted C–space, for example with $w(x) = (1 - x^2)^\alpha$, $\alpha > 0$?

References

[1]   N. N. Andreev, V. A. Yudin: *Polynomials of minimum deviation from zero and cubature formulas of Chebyshev type*, Proceedings Steklov Inst., to appear (2001).

[2]   H. Ehlich, K. Zeller: *Čebyšev–Polynome in mehreren Veränderlichen*, Math. Z. **93** (1966), 142–143.

[3]   W. B. Gearhart: *Some Chebyshev approximations by polynomials of two variables*, J. Approx. Theory **8** (1973), 195–209.

[4]   M. Reimer: *Constructive theory of multivariate functions with an application to tomography*, Wissenschaftsverlag, Mannheim–Wien–Zürich 1990.

[5]   M. Reimer: *Spherical polynomial approximations. A Survey*, Advances in Multivariate Approximation, Math. Research **107**, W. Haußmann, K. Jetter, M. Reimer (eds.), Wiley–VCH, Berlin 1999, pp. 231–252.

[6]   J. M. Sloss: *Chebyshev approximation to zero*, Pacific J. Math. **15** (1965), 305–313.

Addresses:

Nikolay N. Andreev
Department of Function Theory
Steklov Institute of Mathematics
stz. Gubkina 8
Moscow GSP–1
Russia

Vladimir A. Yudin
Department of Higher Mathematics
Moscow Institute of Power Engineering
Krasnokazarmennaya 15
Moscow
Russia

International Series of Numerical Mathematics
Vol. 137, ©2001 Birkhäuser Verlag Basel/Switzerland

The Sign of a Harmonic Function Near a Zero

David H. Armitage and Stephen J. Gardiner

We fix a point x_0 in $\mathbb{R}^n$, where $n \geq 2$, and define

$$\mathcal{H} := \{h : h \text{ is a harmonic function on some ball centered at } x_0$$
$$\text{such that } h(x_0) = 0 \text{ and } h \not\equiv 0\}.$$

For a function h in $\mathcal{H}$ and for all sufficiently small positive numbers r we write

$$E_+(h,r) := \{x \in B(r) : h(x) > 0\},$$

where $B(r)$ denotes the open ball of centre x_0 and radius r. By the maximum principle, $E_+(h,r)$ is never empty. In fact $E_+(h,r)$ is large in the following sense. Writing λ for n–dimensional Lebesgue measure and defining

$$D_+(h,r) := \lambda(E_+(h,r))/\lambda(B(r)),$$

we easily find (see below) that the limit

$$L_+(h) := \lim_{r \to 0+} D_+(h,r)$$

exists and is strictly positive for every h in $\mathcal{H}$. We ask whether more is true.

Question: *Does there exist a positive constant $C(n)$, depending only on the dimension n, such that $L_+(h) \geq C(n)$ for every h in $\mathcal{H}$? If so, what is the largest possible value for $C(n)$?*

This question arose in connection with best one–sided harmonic approximation in the L^1–norm (see [1, Section 4.3]). It may be of some independent interest, since an affirmative answer would yield a quantitative maximum principle for harmonic functions; namely, we would be able to say that if $\liminf_{r \to 0+} D_+(h,r) < C(n)$, then $h = 0$ on a neighbourhood of x_0.

We make the following observations, which are essentially contained in [1]. Without loss of generality, we take x_0 to be the origin 0 of $\mathbb{R}^n$ and consider a function h in $\mathcal{H}$. Then we can write h on a neighbourhood of 0 as the sum of a locally uniformly convergent series $\sum_{j=m}^{\infty} H_j$, where H_j is a homogeneous harmonic polynomial of degree $j \geq 1$ and $H_m \not\equiv 0$. Hence $h(x) = H_m(x) + O(||x||^{m+1})$ as $x \to 0$, and it follows that $D_+(h,r)/D_+(H_m,r) \to 1$ as $r \to 0+$. Since H_m is homogeneous, $D_+(H_m,r)$ is independent of r. Also, the value of $D_+(H_m,r)$ lies strictly between 0 and 1, since H_m takes both positive values and negative values. These remarks show that $L_+(h)$ exists and $0 < L_+(h) = L_+(H_m) < 1$. Thus in our Question we can replace $\mathcal{H}$ by the class of all non–constant homogeneous polynomials on $\mathbb{R}^n$. In the case $n = 2$ any such polynomial H is given in polar coordinates by $H(\rho,\theta) = a\rho^m \cos(m\theta + \alpha)$ for some positive integer m and some constants a and α with $a \neq 0$, and so $D_+(H,r) = 1/2$ for all $r > 0$. This shows that for $n = 2$ our Question is answered with $C(2) = 1/2$, and indeed we have $L_+(h) = 1/2$ for every h in $\mathcal{H}$. (This observation appears as Lemma 1 in [1].) To see that this simple result does not extend to higher dimensions, we note that if H is the harmonic polynomial defined on $\mathbb{R}^3$ by $H(x_1, x_2, x_3) = 2x_1^2 - x_2^2 - x_3^2$, then $D_+(H,r) = 1 - 3^{-1/2}$, as can be shown by a short calculation.

Reference

[1]   D. H. Armitage, S. J. Gardiner: *Best one–sided L^1–approximation by harmonic and subharmonic functions,* in: Advances in Multivariate Approximation, W. Haußmann, K. Jetter, M. Reimer (eds.), Wiley–VCH, Berlin 1999, pp. 43–56.

Addresses:

David H. Armitage
Department of Pure Mathematics
Queen's University Belfast
Belfast BT7 1NN
Northern Ireland

Stephen J. Gardiner
Department of Mathematics
University College Dublin
Dublin 4
Ireland

International Series of Numerical Mathematics
Vol. 137, ©2001 Birkhäuser Verlag Basel/Switzerland

Radial Basis Functions for the Sphere

Brad J. C. Baxter and Simon Hubbert

Abstract

In this paper we compute the ultraspherical series expansions for the more commonly used radial basis functions. In several special cases we provide asymptotic estimates for the decay rate of the coefficients involved. Knowledge of the decay of these coefficients is useful because they enable error estimates for spherical interpolation.

1 Introduction

The multivariate interpolation problem is as follows. Given values $\{f_i\}_{i=1}^N$ of a function $f : \mathbb{R}^d \to \mathbb{R}$ at distinct locations (nodes) $\{x_i\}_{i=1}^N$ in $\mathbb{R}^d$, find an *interpolant* $s : \mathbb{R}^d \to \mathbb{R}$, in a suitable linear space of functions T (the interpolation space), satisfying

$$s(x_i) = f_i, \quad 1 \le i \le N. \tag{1.1}$$

1.1 Radial Basis Function Method

One of the most promising ways of solving this problem is to employ the *Radial Basis Function* (RBF) method. This method specifies the interpolation space

$$T_\phi = \text{span } \{\phi(d(\cdot, x_1)), \ldots, \phi(d(\cdot, x_N))\}, \tag{1.2}$$

where $d(x, y) = \|x - y\|$, $\|\cdot\|$ usually being the Euclidean norm (other norms have been considered; see, for example, [2] and [3]), and $\phi : [0, \infty) \to \mathbb{R}$ is the *radial basis function*.

Now posing the interpolation problem in T_ϕ amounts to finding a function of the form

$$s(x) = \sum_{j=1}^N \lambda_j \phi(d(x, x_j)), \qquad \text{for } \lambda_j \in \mathbb{R}, \quad 1 \le j \le N,$$

satisfying conditions (1.1). This is equivalent to solving the following linear system:

$$A\lambda = f, \qquad (1.3)$$

where $A \in \mathbb{R}^{N \times N}$ is defined by

$$A_{i,j} = \phi(d(x_i, x_j)), \qquad 1 \le i, j \le N. \qquad (1.4)$$

Thus a unique interpolant $s \in T_\phi$ exists for any f if and only if the interpolation matrix A is non–singular.

Definition 1.1. *A function $\phi : [0, \infty) \to \mathbb{R}$ is said to be:*

(i) *Strictly positive definite (SPD) on $\mathbb{R}^d$ whenever its associated interpolation matrix (1.4) is positive definite on $\mathbb{R}^N$, for all distinct $\{x_i\}_{i=1}^N$ in $\mathbb{R}^d$.*

(ii) *Conditionally strictly positive definite of order m ($CSPD(m)$) on $\mathbb{R}^d$ whenever its associated interpolation matrix (1.4) is positive definite on the subspace of $\mathbb{R}^N$ defined by*

$$V_{m-1} = \{\lambda = (\lambda_1, \ldots \lambda_N)^T \in \mathbb{R}^N : \sum_{i=1}^N \lambda_i p(x_i) = 0 \text{ for all } p \in \Pi_{m-1}(\mathbb{R}^d)\},$$

for all distinct $\{x_i\}_{i=1}^N$ in $\mathbb{R}^d$. Here $\Pi_{m-1}(\mathbb{R}^d)$ denotes the space of all d–variate polynomials of degree at most $m - 1$.

If ϕ is SPD on $\mathbb{R}^d$, then there exists a unique interpolant $s \in T_\phi$ since the interpolation matrix is, by definition, positive definite and hence non–singular. If ϕ is $CSPD(m)$ on $\mathbb{R}^d$ however, it can be shown that if the interpolation n-odes are $\Pi_{m-1}(\mathbb{R}^d)$–unisolvent — the only element of $\Pi_{m-1}(\mathbb{R}^d)$ that vanishes at every node is the zero polynomial — then there exists a unique interpolant $s \in T_\phi \oplus \Pi_{m-1}(\mathbb{R}^d)$, that is

$$s(x) = \sum_{j=1}^N \lambda_j \phi(d(x, x_j)) + p(x),$$

where $\lambda = (\lambda_1, \ldots \lambda_N)^T \in V_{m-1}$ and $p \in \Pi_{m-1}(\mathbb{R}^d)$; see [13] for details.

The functions used in the RBF method are usually either SPD or $CSPD(m)$ on $\mathbb{R}^d$. The following is a list of the more common examples.

Gaussian : $\phi(r) = e^{-\alpha r^2}$, $\alpha > 0$;

Potential Spline : $\phi(r) = (-1)^{\lfloor \beta \rfloor + 1} r^{2\beta}$, $\beta > 0$ and $\beta \notin \mathbb{Z}_+ = \{1, 2, \ldots\}$;

Thin Plate Spline : $\phi(r) = (-1)^{k+1} r^{2k} \log(r)$, $k \in \mathbb{Z}_+$;

Multiquadric : $\phi(r) = (-1)^{[\beta]+1}(r^2 + c^2)^\beta$, $\beta > 0$, $\beta \notin \mathbb{Z}_+$, and $c > 0$;

Inverse Multiquadric : $\phi(r) = (r^2 + c^2)^\beta$, $-d/2 < \beta < 0$, $\beta \notin \mathbb{Z}$, and $c > 0$.

1.2   Zonal Basis Function Method

The RBF method can be specialised if attention is turned to the case where the distinct locations $x_1, \ldots x_N$ are known to lie on the unit sphere S^{d-1} in $\mathbb{R}^d$, $d \geq 2$. To transfer the method we consider the interpolation space

$$T_\psi = \text{span } \{\psi(g(\cdot, x_1)), \ldots, \psi(g(\cdot, x_N))\}, \qquad (1.5)$$

where $g(x, y) = \arccos(x^T y)$ denotes the geodesic metric, and $\psi : [0, \pi] \to \mathbb{R}$ is called a *zonal basis function* (ZBF).

Following the development of the RBF method, it is clear that interpolation is unique in T_ψ if and only if the associated interpolation matrix $B \in \mathbb{R}^{N \times N}$ defined by

$$B_{i,j} = \psi(g(x_i, x_j)), \qquad 1 \leq i, j \leq N, \qquad (1.6)$$

is non–singular.

Definition 1.2. *A function $\psi : [0, \pi] \to \mathbb{R}$ is said to be:*

(i) *Strictly positive definite (SPD) on S^{d-1} whenever its associated interpolation matrix (1.6) is positive definite on $\mathbb{R}^N$, for all distinct $\{x_i\}_{i=1}^N$ on S^{d-1}.*

(ii) *Conditionally strictly positive definite of order m (CSPD(m)) on S^{d-1} whenever its associated interpolation matrix (1.6) is positive definite on the subspace of $\mathbb{R}^N$ given by*

$$W_{m-1} = \{\lambda = (\lambda_1, \ldots \lambda_N)^T \in \mathbb{R}^N : \sum_{i=1}^N \lambda_i Y(x_i) = 0 \text{ for all } Y \in H_{m-1}(S^{d-1})\},$$

for all distinct $\{x_i\}_{i=1}^N$ on S^{d-1}. Here $H_{m-1}(S^{d-1})$ denotes the space of all spherical harmonics on S^{d-1} of order at most $m - 1$.

With Definition 1.2, the specialisation of the RBF method to the sphere (*the ZBF method*) is complete. In particular, interpolation is unique in T_ψ if ψ is *SPD* on S^{d-1}. If ψ is *CSPD(m)* on S^{d-1} however, it can be shown that if the interpolation nodes are $H_{m-1}(S^{d-1})$–unisolvent — the only element of

$H_{m-1}(S^{d-1})$ that vanishes at every node is the zero spherical harmonic — then there exists a unique interpolant $s \in T_\psi \oplus H_{m-1}(S^{d-1})$, that is

$$s(x) = \sum_{j=1}^{N} \lambda_j \psi(g(x, x_j)) + Y(x),$$

where $\lambda = (\lambda_1, \dots \lambda_N)^T \in W_{m-1}$ and $Y \in H_{m-1}(S^{d-1})$; see [5] for details. We remark that the role of the spherical harmonic space $H_{m-1}(S^{d-1})$ within the ZBF method is equivalent to the role of the polynomial space $\Pi_{m-1}(\mathbb{R}^d)$ within the RBF method, indeed $H_{m-1}(S^{d-1}) = \Pi_{m-1}(\mathbb{R}^d)|_{S^{d-1}}$; (see [11] or [14]).

Using the work of Schoenberg [16], and extensions thereof [5], we can formulate the following theorem.

Theorem 1.3. *If ψ is $CSPD(m)$ on S^{d-1}, then ψ has the following form*

$$\psi(\theta) = \sum_{k=0}^{\infty} a_k P_k^\lambda(\cos(\theta)), \tag{1.7}$$

where

$$a_k \geq 0 \quad for \quad k \geq m \quad and \quad \sum_{k=0}^{\infty} a_k P_k^\lambda(1) < \infty. \tag{1.8}$$

Here $\{P_k^\lambda\}$ denote the ultraspherical polynomials ([1], 22.2.3) and $\lambda = (d-2)/2$.

Remarks 1.4. (i) *The case $\psi \in SPD(m)$ is covered by setting $m = 0$ in Theorem 1.3.*

(ii) *In [12], a framework is established for solving the interpolation problem on a compact Riemannian manifold $\mathbf{M}$ using SPD kernels $\kappa : \mathbf{M} \times \mathbf{M} \to \mathbb{R}$. The ZBF method with ψ SPD is a specific instance of this more general approach for $\mathbf{M} = S^{d-1}$.*

(iii) *In view of Theorem 1.3 we choose to consider each zonal function ψ as a function of the inner product, $x^T y$, since $\cos(g(x, y)) = x^T y$.*

The complete characterization of the class of functions of the form (1.7) satisfying (1.8) that are $CSPD(m)$ on S^{d-1} remains an open problem. Several researchers have investigated this in recent papers; in particular, in [18], it is shown that a sufficient condition is $a_k > 0$, for $k \geq m$. (See [15] for an extension of this work). One can use this condition to generate candidate zonal functions to be used within the ZBF method. The following is a list of functions ψ which are SPD on S^2 for example:

$\psi(t) = (1 + h^2 - 2ht)^{-1/2}$, where $a_k = h^k$, for $0 < h < 1$;

$\psi(t) = (1 - h^2)(1 + h^2 - 2ht)^{-3/2}$, where $a_k = (2k + 1)h^k$, for $0 < h < 1$;

$\psi(t) = 1 - \sqrt{\frac{1-t}{2}}$, where $a_0 = 1/3$ and $a_k = \frac{2}{(2k-1)(2k+3)}$, $k \geq 1$.

2 Radial Functions for Spheres

Most of the recent research regarding the ZBF method is of a theoretical nature, and very little has been reported of its performance in practice (see, however [6]). Much more is known about the RBF method and so a potential user may wish to take a common radial function and use it as a zonal basis function; indeed a radial function ϕ that is $CSPD(m)$ on $\mathbb{R}^d$ is also $CSPD(m)$ on S^{d-1}. Furthermore, the RBFs remain well defined if the interpolation problem is set on a perturbed sphere, which is likely to be the case for several practical applications.

In order to take advantage of the existant ZBF theory (especially convergence results [7] and [10]), it is desirable to have the ultraspherical series expansions (1.7) for all the common radial functions. The remainder of this section addresses precisely this issue. In order to use radial functions on the sphere one usually employs

$$d(x, y) = \|x - y\| = \sqrt{2 - 2x^T y}, \qquad x, y \in S^{d-1}. \tag{2.1}$$

In particular, if ϕ is $CSPD(m)$ on $\mathbb{R}^d$ then the zonal function $\psi(t) = \phi(\sqrt{2 - 2t})$ is $CSPD(m)$ on S^{d-1}, and so, by Theorem (1.1), has an expansion

$$\psi(t) = \sum_{n=0}^{\infty} a_n P_n^\lambda(t), \qquad -1 \leq t \leq 1,$$

where the coefficients (a_n) satisfy (1.8). The ultraspherical polynomials P_n^λ are given by Rodrigues' formula ([1], 22.11.2)

$$P_n^\lambda(t) = c_n(\lambda)(1 - t^2)^{1/2-\lambda} \frac{d^n}{dt^n} (1 - t^2)^{n+\lambda-1/2}, \quad n \in \mathbb{N} = \{0, 1, \ldots\}, \tag{2.2}$$

where

$$c_n(\lambda) = \frac{(-1)^n \pi^{1/2} 2^{(1-n-2\lambda)} \Gamma(n + 2\lambda)}{\Gamma(n + \lambda + \frac{1}{2}) \Gamma(n + 1) \Gamma(\lambda)}. \tag{2.3}$$

We note that these are simply the Legendre polynomials when $\lambda = 1/2$. They satisfy the orthogonality relation ([1], 22.2.3)

$$\int_{-1}^{1} P_m^\lambda(t) P_n^\lambda(t)(1-t^2)^{\lambda-1/2} dt = \begin{cases} 0, & m \neq n, \\ d_n, & m = n, \end{cases} \tag{2.4}$$

where

$$d_n = \frac{\pi \Gamma(n+2\lambda)}{2^{2\lambda-1}(n+\lambda)\Gamma(n+1)\Gamma(\lambda)^2}, \tag{2.5}$$

and thus the series coefficients are given by

$$a_n = \frac{1}{d_n} \int_{-1}^{1} \psi(t)(1-t^2)^{\lambda-1/2} P_n^\lambda(t) dt, \quad n \in \mathbb{N}. \tag{2.6}$$

Employing (2.2) and integrating by parts n times gives

$$a_n = \frac{(-1)^n c_n(\lambda)}{d_n} \int_{-1}^{1} \psi^{(n)}(t)(1-t^2)^{n+\lambda-1/2} dt. \tag{2.7}$$

2.1   Multiquadrics

Here we consider the function $\phi(r) = (r^2 + c^2)^\beta$, $c > 0$, where $\beta \in \mathbb{R}\backslash\mathbb{Z}$. It is known that ϕ is SPD for $-d/2 < \beta < 0$, and $(-1)^{[\beta]+1}\phi$ is $CSPD([\beta]+1)$ for $\beta > 0$ (see [13]). To use the multiquadric on the sphere we consider $\psi(t) = (2 + c^2 - 2t)^\beta$. Applying (2.7) for $n \in \mathbb{N}$ gives

$$a_n(\beta, \lambda) = \frac{2^{\beta-n}\Gamma(\beta+1)}{\Gamma(\beta-n+1)} \frac{c_n(\lambda)}{d_n} \int_{-1}^{1} (1 + \frac{c^2}{2} - t)^{\beta-n}(1-t^2)^{n+\lambda-1/2} dt.$$

Setting $A = \frac{c^2}{2}$ and $u = \frac{t+1}{2}$, we find

$$a_n(\beta, \lambda) \quad =$$

$$\frac{2^{2n+2\lambda+\beta}(2+A)^{\beta-n}\Gamma(\beta+1)}{\Gamma(\beta-n+1)} \frac{c_n(\lambda)}{d_n} \int_{0}^{1} (1 - \frac{2u}{2+A})^{\beta-n}(1-u)^{n+\lambda-1/2} u^{n+\lambda-1/2} du.$$

Using the identity (see [1], 15.3.1)

$$\int\limits_0^1 (1 - zu)^{-a}(1 - u)^{c-b-1}u^{b-1}\,du = \frac{\Gamma(b)\Gamma(c - b)}{\Gamma(c)}F(a, b; c; z),$$

with $a = n - \beta$, $b = n + \lambda + 1/2$, $c = 2b$ and $z = \frac{2}{2+A}$, we see that

$$a_n(\beta, \lambda) = \alpha_n(\beta, \lambda)F(n - \beta, n + \lambda + 1/2; 2(n + \lambda + 1/2); \frac{2}{2 + A}), \qquad (2.8)$$

where

$$\alpha_n(\beta, \lambda) = \frac{2^{2n+2\lambda+\beta}(2 + A)^{\beta-n}\Gamma(\beta + 1)\Gamma(n + \lambda + \frac{1}{2})^2}{\Gamma(\beta - n + 1)\Gamma(2(n + \lambda + \frac{1}{2}))}\frac{c_n(\lambda)}{d_n} \qquad (2.9)$$

which on substituting (2.3) and (2.5)

$$= \frac{(-1)^n 2^{n+2\lambda+\beta}(2 + A)^{\beta-n}(n + \lambda)\Gamma(\lambda)\Gamma(\beta + 1)}{\Gamma(\beta - n + 1)\pi^{1/2}}\frac{\Gamma(n + \lambda + \frac{1}{2})}{\Gamma(2(n + \lambda + \frac{1}{2}))} \qquad (2.10)$$

and $F(a, b; c; z)$ is the Gauss Hypergeometric series (see [1], 15.1.1) defined by

$$F(a, b; c; z) = \frac{\Gamma(c)}{\Gamma(a)\Gamma(b)}\sum_{k=0}^{\infty}\frac{\Gamma(a + k)\Gamma(b + k)}{\Gamma(c + k)}\frac{z^k}{k!}. \qquad (2.11)$$

This series is absolutely convergent for $|z| \leq 1$ provided the real part $\mathcal{R}(c-a-b)$ is positive, that is, $-d/2 < \beta$. Thus (2.8) holds for all multiquadrics.

2.2 Potential Splines

Here we consider the function $\phi(r) = r^{2\beta}$, for $\beta > 0$ and $\beta \notin \mathbb{Z}_+$. It is known that $(-1)^{\lfloor\beta\rfloor+1}\phi$ is $CSPD([\beta]+1)$ (see [13]). To use the potential splines on the sphere we consider, $\psi(t) = (2-2t)^{\beta}$. This can be derived from the multiquadric case above by simply setting $A = \frac{c^2}{2} = 0$, i.e. $\frac{2}{2+A} = 1$. Using the results from Section 2.1 and the following identity ([1], 15.1.20)

$$F(a, b; c; 1) = \frac{\Gamma(c)\Gamma(c - a - b)}{\Gamma(c - a)\Gamma(c - b)},$$

we can deduce the ultraspherical coefficients of the potential splines

$$a_n(\beta, \lambda) = (-1)^n\pi^{-1/2}2^{2\lambda}(n + \lambda)\Gamma(\lambda)\frac{2^{2\beta}\Gamma(\beta + 1)\Gamma(\beta + \lambda + \frac{1}{2})}{\Gamma(\beta + 1 - n)\Gamma(\beta + n + 1 + 2\lambda)}. \qquad (2.12)$$

2.3   Thin Plate Splines

Here we consider the function $\phi(r) = r^{2k} \log r$, for $k \in \mathbb{Z}_+$. It is known that $(-1)^{k+1}\phi$ is $CSPD(k+1)$ (see [13]). To use the thin plate splines on the sphere we consider, $\psi(t) = \frac{1}{2}(2 - 2t)^k \log(2 - 2t)$. This function can be derived from the potential spline using the observation

$$\psi(t) = \frac{1}{2}\frac{\partial}{\partial\beta}(2 - 2t)^{\beta}\Big|_{\beta=k}. \tag{2.13}$$

Thus the ultraspherical coefficients of the thin plate splines $b_n(k, \lambda)$ are given by

$$b_n(k, \lambda) = \frac{1}{2}\frac{\partial}{\partial\beta}a_n(\beta, \lambda)\Big|_{\beta=k}, \tag{2.14}$$

where $a_n(\beta, \lambda)$ are as in (2.12). In particular, we rewrite (2.12) as

$$a_n(\beta, \lambda) = (-1)^n \pi^{-1/2} 2^{2\lambda}(n + \lambda)\Gamma(\lambda)h(\beta), \tag{2.15}$$

where

$$h(\beta) = \frac{2^{2\beta}\Gamma(\beta + 1)\Gamma(\beta + \lambda + \frac{1}{2})}{\Gamma(\beta + 1 - n)\Gamma(\beta + n + 1 + 2\lambda)}. \tag{2.16}$$

In order to differentiate $h(\beta)$, we consider the so called digamma function ([1], 6.3.1), which is defined by $\Psi(z) = \Gamma'(z)/\Gamma(z)$ for $z \neq 0, -1, -2, \ldots$. Then for $\beta = k \geq n$

$$h'(k) = h(k)\{\Psi(k + 1) + \Psi(k + \lambda + \frac{1}{2}) + 2\log 2 \tag{2.17}$$

$$-\Psi(k + 1 - n) - \Psi(k + n + 2\lambda + 1)\}$$

We can also write $\Gamma(\beta + 1) = \beta(\beta - 1)\cdots(\beta - n + 1)\Gamma(\beta - n + 1)$ and so consider $h(\beta)$ as

$$h(\beta) = \frac{2^{2\beta}\Gamma(\beta + \lambda + \frac{1}{2})\beta(\beta - 1)\cdots(\beta - n + 1)}{\Gamma(\beta + n + 2\lambda + 1)} = \frac{u(\beta)v(\beta)}{w(\beta)},$$

where $u(\beta) = 2^{2\beta}\Gamma(\beta + \lambda + 1/2)$, $v(\beta) = \beta(\beta - 1)\cdots(\beta - n + 1)$ and $w(\beta) = \Gamma(\beta + n + 2\lambda + 1)$. Thus

$$h'(k) = \frac{w(k)\{u'(k)v(k) + u(k)v'(k)\} - u(k)v(k)w'(k)}{w(k)^2}$$

and, since $v(k) = 0$, for all $k < n$, this is simply

$$h'(k) = \frac{u(k)v'(k)}{w(k)}$$

Furthermore $v'(k) = (-1)^{n-(k+1)}\Gamma(k+1)\Gamma(n-k)$, from which we can see

$$h'(k) = \frac{(-1)^{n-(k+1)}2^{2k}\Gamma(k+\lambda+\frac{1}{2})\Gamma(k+1)\Gamma(n-k)}{\Gamma(k+n+2\lambda+1)}. \quad (2.18)$$

We can now use equations (2.17) and (2.18) to deduce the ultraspherical coefficients for the thin plate spline; for $k \geq n$

$$b_n(k,\lambda) = a_n(k,\lambda)\{\Psi(k+1) + \Psi(k+\lambda+\frac{1}{2}) + 2\log 2 - \Psi(k+1-n) \quad (2.19)$$

$$-\Psi(k+n+2\lambda+1)\};$$

whilst for $k < n$

$$b_n(k,\lambda) = \frac{(-1)^{k+1}2^{2(k+\lambda)}(n+\lambda)\Gamma(\lambda)\Gamma(k+\lambda+\frac{1}{2})\Gamma(k+1)\Gamma(n-k)}{2\pi^{1/2}\Gamma(k+n+1+2\lambda)}. \quad (2.20)$$

2.4 Gaussians

Here we consider the function $\phi(r) = e^{-\alpha r^2}$, for $\alpha > 0$. It is well known that ϕ is SPD (see [13]). To use the Gaussian on the sphere we consider, $\psi(t) = e^{-2\alpha}e^{2\alpha t}$. Again we apply formula (2.7) to obtain

$$a_n(\alpha,\lambda) = \frac{(-1)^n e^{-2\alpha}(2\alpha)^n c_n(\lambda)}{d_n} \int_{-1}^{1} e^{2\alpha t}(1-t^2)^{n+\lambda-1/2}dt.$$

The integral in the above formula represents the modified Bessel function $I_{n+\lambda}$; specifically we have ([17], 3.71)

$$I_{n+\lambda}(2\alpha) = \frac{\alpha^{n+\lambda}}{\Gamma(n+\lambda+\frac{1}{2})\pi^{1/2}} \int_{-1}^{1} e^{2\alpha t}(1-t^2)^{n+\lambda-1/2}dt. \quad (2.21)$$

Therefore we deduce the ultraspherical coefficients

$$a_n(\alpha, \lambda) = \frac{(-1)^n \pi^{1/2} 2^n e^{-2\alpha} \Gamma(n + \lambda + \frac{1}{2}) \, c_n(\lambda)}{\alpha^\lambda} \frac{c_n(\lambda)}{d_n} I_{n+\lambda}(2\alpha), \quad n \in \mathbb{N},$$

and, on substituting (2.3) and (2.5),

$$a_n(\alpha, \lambda) = \frac{(n + \lambda)\Gamma(\lambda)e^{-2\alpha}}{\alpha^\lambda} I_{n+\lambda}(2\alpha). \tag{2.22}$$

3 Common Radial Basis Functions for the 2–Sphere

In this concluding section we specialise the results of Section 2 to the sphere S^2, in which case $\lambda = 1/2$. Furthermore we apply the results to the radial functions in their more familiar form.

3.1 The Inverse Multiquadric: $\phi(r) = (r^2 + c^2)^{-1/2}$.

Here we apply (2.8) with $\lambda = 1/2$ and $\beta = -1/2$ giving, for $n \in \mathbb{N}$,

$$a_n = a_n(-1/2, 1/2) = \alpha_n(-1/2, 1/2) F(n + 1/2, n + 1; 2(n + 1); \frac{4}{4 + c^2}).$$

Considering the following identity ([1], 15.1.13):

$$F(a, \frac{1}{2} + a; 1 + 2a; z) = 2^{2a}(1 + \sqrt{1 - z})^{-2a}$$

setting $a = n + \frac{1}{2}$ and $z = \frac{4}{4+c^2}$ allows us to deduce:

$$a_n = \frac{(n + \frac{1}{2})(-1)^n \pi}{\Gamma(\frac{1}{2} - n)\Gamma(n + \frac{3}{2})} \left(\frac{2}{c + \sqrt{4 + c^2}}\right)^{2n+1}, \tag{3.1}$$

this can be simplified further, using the identity ([1], 6.1.17)

$$\Gamma(z)\Gamma(1 - z) = \frac{\pi}{\sin \pi z}. \tag{3.2}$$

In particular setting $z = 1/2 - n$ yields

$$\Gamma(1/2 - n)\Gamma(n + 3/2) = (n + \frac{1}{2})\Gamma(1/2 - n)\Gamma(n + 1/2) = (n + \frac{1}{2})(-1)^n \pi$$

giving

$$a_n = h^{2n+1} \tag{3.3}$$

where $h = \frac{2}{c+\sqrt{4+c^2}} < 1$, thus the coefficients decay at an exponential rate.

3.2 The Multiquadric: $\phi(r) = (r^2 + c^2)^{1/2}$.

Here we apply (2.8) again with $\lambda = 1/2$, $\beta = 1/2$, giving for $n \in \mathbb{N}$,

$$a_n = a_n(1/2, 1/2) = \alpha_n(1/2, 1/2) F(n - 1/2, n + 1; 2(n + 1); \frac{4}{4 + c^2}).$$

A closed form representation for $F(n - 1/2, n + 1; 2(n+1); \frac{4}{4+c^2})$ can be derived quite easily ([1] Section 15). In particular, we have:

$$F(n - 1/2, n+1; 2(n+1); z) = \frac{(n + \frac{1}{2})\sqrt{1-z} + (1 - \frac{z}{2})}{(n + \frac{3}{2})} 2^{2n+1}(1+\sqrt{1-z})^{-(2n+1)}.$$

Setting $z = \frac{4}{4+c^2}$ and multiplying by $\alpha_n(1/2, 1/2)$ gives

$$a_n = \frac{(-1)^n \pi (n + \frac{1}{2})(2 + c^2 + (n + \frac{1}{2})c\sqrt{4 + c^2})}{2\Gamma(\frac{3}{2} - n)\Gamma(n + \frac{5}{2})} \left(\frac{2}{c + \sqrt{4 + c^2}}\right)^{2n+1}. \qquad (3.4)$$

Further simplification is possible, setting $z = 3/2 - n$ in (3.2) gives

$$\Gamma(3/2 - n)\Gamma(n + 5/2) = (n + \frac{3}{2})(n + \frac{1}{2})(n - \frac{1}{2})\Gamma(3/2 - n)\Gamma(n - 1/2)$$

$$= (n + \frac{3}{2})(n + \frac{1}{2})(n - \frac{1}{2})(-1)^{n-1}\pi,$$

thus

$$a_n = -\frac{(2 + c^2 + (n + \frac{1}{2})c\sqrt{4 + c^2})}{2(n + \frac{3}{2})(n - \frac{1}{2})} h^{2n+1} = O\left(\frac{h^{2n+1}}{n}\right). \qquad (3.5)$$

3.3 The Pseudo Cubic: $\phi(r) = r^3$

Here we simply set $\beta = 3/2$ and $\lambda = 1/2$ in (2.12) giving, for $n \in \mathbb{N}$,

$$a_n = a_n(3/2, 1/2) = (-1)^n 2^4 \frac{(n + \frac{1}{2})\Gamma(\frac{5}{2})^2}{\Gamma(\frac{5}{2} - n)\Gamma(n + \frac{7}{2})} \qquad (3.6)$$

Simplification is again possible, setting $z = 5/2 - n$ in (3.2) gives

$$\Gamma(5/2 - n)\Gamma(n + 7/2) = (n + \frac{5}{2})(n + \frac{3}{2})(n + \frac{1}{2})(n - \frac{1}{2})(n - \frac{3}{2})(-1)^n\pi,$$

thus

$$a_n = \frac{9}{(n + \frac{5}{2})(n + \frac{3}{2})(n - \frac{1}{2})(n - \frac{3}{2})} = O\left(\frac{1}{n^4}\right). \qquad (3.7)$$

3.4   The Thin Plate Spline: $\phi(r) = r^2 \log r$

Here we simply set $k = 1$ and $\lambda = 1/2$ in (2.20), for $n > 1$ this provides

$$a_n = b_n(1, 1/2) = 4(n + 1/2)\frac{\Gamma(2)^2\Gamma(n - 1)}{\Gamma(n + 3)}$$

$$= \frac{4(n + \frac{1}{2})}{(n + 2)(n + 1)n(n - 1)} = O\left(\frac{1}{n^3}\right). \qquad (3.8)$$

3.5   The Gaussian: $\phi(r) = e^{-\alpha r^2}$

Setting $\lambda = 1/2$ in (2.22) yields, for $n \in \mathbb{N}$,

$$a_n = a_n(\alpha, 1/2) = \sqrt{\frac{\pi}{\alpha}}(n + 1/2)e^{-2\alpha}I_{n+\frac{1}{2}}(2\alpha),$$

employing 2.21 gives

$$a_n = \frac{(n + \frac{1}{2})e^{-2\alpha}\alpha^n}{\Gamma(n + 1)} \int_{-1}^{1} e^{2\alpha t}(1 - t^2)^n dt. \qquad (3.9)$$

We can derive the asymptotic behaviour using the well–known method of Laplace. However, we prefer a direct approach, which we present for the convenience of the reader. Consider the integral appearing in (3.9), that is

$$G_n = \int_{-1}^{1} e^{2\alpha t}(1 - t^2)^n dt. \qquad (3.10)$$

Setting $\tau = \sqrt{n}t$, we obtain

$$\sqrt{n}G_n = \int_{-\infty}^{\infty} f_n(\tau)d\tau$$

where

$$f_n(\tau) = \begin{cases} e^{\frac{2\alpha\tau}{\sqrt{n}}}(1 - \tau^2/n)^n, & |\tau| \le \sqrt{n} \\ 0, & |\tau| > \sqrt{n}. \end{cases} \tag{3.11}$$

Observing that $0 \le f_n(\tau) \le e^{2\alpha}e^{-\tau^2}$ and $\lim_{n \to \infty} f_n(\tau) = e^{-\tau^2}$ allows us to employ the dominated convergence theorem,

$$\sqrt{n}G_n = \int_{-\infty}^{\infty} f_n(\tau)d\tau \to \int_{-\infty}^{\infty} e^{-\tau^2}d\tau = \sqrt{\pi}, \quad as \quad n \to \infty. \tag{3.12}$$

Also we have Stirling's formula ([1], 6.1.38)

$$\Gamma(n + 1) = n! \sim \sqrt{2\pi}n^{n+1/2}e^{-n}, \quad as \quad n \to \infty. \tag{3.13}$$

Employing (3.12) and (3.13) together in (3.9) gives

$$a_n \sim \frac{e^{-2\alpha}}{2^{1/2}}\left(\frac{e\alpha}{n}\right)^n, \quad as \quad n \to \infty,$$

i.e. the Gaussian coefficients decay at an exponential rate.

3.6 Concluding Remarks

The motivation for this work stems from recent results on error estimates for spherical interpolation ([4], [7] and [10]). These topics have been investigated by several mathematicians, in particular the Leicester group ([8] and [9]). Specifically the report [8] calculates the ultraspherical coefficients for the Duchon splines which are also contained in this paper, the approach taken (private communication) however is quite distinct from the one given here.

For practical purposes a potential user would prefer to work with a basis function ψ with a closed form representation and with provably good approximation properties. The results of this paper allow us to provide convergence results for the common RBF's restricted to the sphere.

The process of restricting radial functions to the sphere clearly provides suitable ZBF's. However the class of *all* suitable ZBF's is much larger, containing, in addition, the *truly zonal* functions, i.e. those that are SPD or $CSPD(m)$ on S^{d-1} but not on $\mathbb{R}^d$. It is not clear whether choosing a truly zonal function provides any advantages over a restricted RBF, and this is an obvious topic for further research.

References

[1]   M. Abramowitz, I. A. Stegun: *Handbook of Mathematical Functions*, National Bureau of Standards, Dover 1964.

[2]   B. J. C. Baxter: *Conditionally positive definite functions and p-norm distance matrices*, Constr. Approx. **7** (1991), 427–440.

[3]   N. Dyn, W. A. Light, E. W. Cheney: *Interpolation by piecewise linear radial basis functions*, J. Approx. Theory. **59** (1989), 202–223.

[4]   N. Dyn, F. Narcowich, J. Ward: *Variational principles and Sobolev–type esimates for generalized interpolation on a Riemannian manifold*, Constr. Approx. **15** (1999), 175–208.

[5]   W. Freeden, T. Gervens, M. Schreiner: *Constructive Approximation on the Sphere With Applications to Geomathematics*, Oxford University Press, Oxford 1998.

[6]   S. Hubbert: *Spherical interpolation using basis function methods*, preprint (2000).

[7]   K. Jetter, J. Stöckler, J. Ward: *Error estimates for scattered data interpolation on spheres*, Math. Comp. **68** (1999), 733–747.

[8]   J. Levesley: *Convergence of Euclidean radial basis approximation on spheres*, preprint, 2000.

[9]   J. Levesley, W. A. Light, D. Ragozin, X. Sun: *A simple approach to the variational theory for interpolation on spheres*, in: New Developments in Approximation Theory, M. D. Buhmann, M. Felten, D. Mache, M. W. Müller (eds.), Birkhäuser, Basel 1998, pp. 117–143.

[10]   T. M. Morton, M. Neamtu: *Error bounds for solving pseudodif-
         ferential equations on spheres by collocation using zonal kernels,*
         preprint, 2000.

[11]   C. Müller: *Spherical Harmonics,* Lect. Notes Math. **17**, Springer,
         Berlin 1966.

[12]   F. Narcowich: *Generalized Hermite interpolation and positive def-
         inite Kernels on a Riemannian manifold,* J. Math. Anal. Appl.
         190 (1995), 165–193.

[13]   M. J. D. Powell: *The theory of radial basis function approximation
         in 1990,* in: Advances in Numerical Analysis II: Wavelets, Sub-
         divisions and Radial Basis Functions, W. A. Light (ed.), Oxford
         University Press, Oxford 1992, pp. 105–210.

[14]   E. M. Stein, G. Weiss: *Introduction to Fourier Analysis on Eu-
         clidean Spaces,* Princeton University Press, Princeton 1971.

[15]   A. Ron, X. Sun: *Strictly positive definite functions on spheres in
         Euclidean spaces,* Math. Comp. **65** (1996), 1513–1530.

[16]   I. J. Schoenberg: *Positive definite functions on spheres,* Duke
         Math. J. **9** (1942), 96–108.

[17]   G. N. Watson: *A Treatise on the Theory of Bessel Functions,*
         Cambridge University Press, Cambridge 1944.

[18]   Y. Xu, E. W. Cheney: *Strictly positive definite functions on
         spheres,* Proc. Amer. Math. Soc. **116** (1992), 977–981.

Address:

Brad J. C. Baxter, Simon Hubbert
Department of Mathematics
Imperial College
London SW7 2BZ
England

International Series of Numerical Mathematics
Vol. 137, ©2001 Birkhäuser Verlag Basel/Switzerland

Cubature Formulae for Polyharmonic Functions

Borislav Bojanov

Abstract

We construct cubature formulae for the ball in $\mathbb{R}^d$, based on integrals of the function $v(x)$, its normal derivatives or iterated Laplacians $\Delta^j v$ $(j = 0, \ldots, m_i - 1)$ over n distinct $(d-1)$–dimensional hyperspheres $S(\rho_i)$ of radius ρ_i, respectively, which integrate exactly wide classes of functions defined as the *harmonic span* of given radial functions, and including the class of polyharmonic functions of fixed order.

1 Introduction

The study of approximation problems dealing with multivariate algebraic polynomials encounter serious technical difficulties. In many cases, even classical univariate results, that have simple elegant proofs, do not admit natural multivariate extensions or the study of their multivariate counterparts is blocked by the lack of adequate technique. A typical example is the interpolation problem based on function values at fixed points.

Recently we exploited an approach based on polyharmonic functions to obtain cubature formulae that are exact for classes of multivariate algebraic polynomials (see [4], [6]). The idea is quite transparent: The space of polyharmonic functions in $\mathbb{R}^d$ of degree m contains the set of algebraic polynomials of d variables of degree $2m - 1$. Thus any approximation rule that is good (or exact) for the polyharmonic functions would be good for the corresponding set of algebraic polynomials. Thus the technique developed in the theory of polyharmonic functions can be used to study polynomial approximation problems. This approach turned out to be very fruitful when studying integration over a ball, because of the extremely nice properties of the polyharmonic functions and their integrals on the ball. We illustrate here our method constructing new cubature formulae of highest algebraic and polyharmonic degree of precision.

We start with some necessary notations and definitions.

For a given $r > 0$, let us denote by $B(r)$ the closed ball in $\mathbb{R}^d$ of radius r and a center at zero, that is,

$$B(r) := \{x = (x_1, \ldots, x_d) \in \mathbb{R}^d : |x| := \left(\sum_{k=1}^{d} x_i^2 \right)^{1/2} \leq r\}.$$

A function u is said to be *polyharmonic* of degree m in a given domain $D \subset \mathbb{R}^d$ if $\Delta^m u = 0$ on D. Here, as usual, Δ^m is the m-th iterate of the Laplace operator

$$\Delta u := \sum_{i=1}^{n} \frac{\partial^2 u}{\partial x_i^2}, \quad \Delta^m u := \Delta(\Delta^{m-1} u).$$

In what follows we shall denote by $H^{(m)}(B(r))$ the space of all functions u that are polyharmonic of degree m in a domain containing $B(r)$.

The next theorem due to Almansi [1] gives an important representation of the polyharmonic functions from $H^{(m)}(B(r))$. The proof can be found also in [14], [3]).

Almansi Theorem. *If* $u \in H^{(m)}(B(r))$, *then there exist unique functions* $h_0(x), h_1(x), \cdots, h_{m-1}(x)$, *each harmonic in* $B(r)$, *such that*

$$u(x) = \sum_{j=0}^{m-1} |x|^{2j} h_j(x) \quad \text{for} \quad x \in B(r). \tag{1}$$

It is seen that any function of form (1) is polyharmonic of degree m. Thus $H^{(m)}(B(r))$ consists of all functions that admit representation of the form (1) with harmonic $\{h_j\}$.

We shall use the following slightly more general version of the Almansi representation which was given in [6].

Denote by $\pi_k(\mathbb{R}^d)$ the set of all algebraic polynomials of d variables of total degree less then or equal to k. In particular, $\pi_k = \pi_k(\mathbb{R})$ is the class of univariate polynomials of degree $\leq k$.

Theorem A. *Let* $\phi_0(t), \cdots, \phi_{m-1}(t)$ *be any basis in the space* π_{m-1}. *If* $u \in H^{(m)}(B(r))$, *then there exist unique functions* $h_0(x), h_1(x), \cdots, h_{m-1}(x)$, *each harmonic in* $B(r)$, *such that*

$$u(x) = \sum_{j=0}^{m-1} \phi_j(|x|^2) h_j(x) \quad \text{for} \quad x \in B(r).$$

Integrals of polyharmonic functions over a ball can be easily reduced to univariate integration. To see this, let us recall first some classical results.

According to the Gauss mean–value theorem, for every harmonic function h on $B(r)$, we have

$$\int_{B(r)} h(x) \, dx \; = \; \text{Vol } B(r) \, h(0),$$

where Vol $B(r) := r^d \, [\pi^{d/2}/\Gamma(d/2+1)]$ is the volume of $B(r)$. The Gauss formula can be considered as a cubature formula for harmonic functions on the ball $B(r)$. As mentioned in [4], this result holds in a more general situation of weighted integration. Namely, the following is true.

Lemma 1. *Let* $\varphi(t)$ *be any integrable function on* $[0, r]$. *Then*

$$\int_{B(r)} \varphi(|x|) h(x) \, dx = h(0) \int_{B(r)} \varphi(|x|) \, dx$$

for every function h *which is harmonic in* $B(r)$.

The proof can be seen in [4].

Recall also another relation which follows from the Gauss mean–value theorem and the Green formula:

$$\int_{S(r)} h(x) \, d\sigma(x) = \gamma_d r^{d-1} h(0), \qquad \gamma_d := d\pi^{d/2}/\Gamma(d/2+1).$$

As usual, here $S(r)$ denotes the sphere of radius r in $\mathbb{R}^d$ (i.e., the boundary of $B(r)$),

$$S(r) := \{x \in \mathbb{R} : |x| = r\}$$

and $d\sigma$ is the $(d-1)$–dimensional surface measure on the sphere $S(r)$.

A direct application of the above formulae yields

$$
\begin{aligned}
\int_{B(r)} \varphi(|x|) h(x) \, dx \; &= \; h(0) \int_{B(r)} \varphi(|x|) \, dx \\
&= \; h(0) \int_0^r \int_{S(t)} \varphi(|x|) \, d\sigma(x) \, dt \\
&= \; h(0) \gamma_d \int_0^r \varphi(t) t^{d-1} \, dt.
\end{aligned}
$$

Therefore, integrals of polyharmonic functions are related to weighted univariate integrals of algebraic polynomials. Moreover, the calculation we just demonstrated, suggests a way for numerical integration of any linear combination of expressions of the form $\varphi_j(|x|)h_j(x)$ with harmonic h_j.

That is why we shall consider here classes of functions introduced in [4] and called the *harmonic span* of a given system $\{\varphi_j\}$. Precisely, assume that $\overline{\varphi} := \{\varphi_0(t), \ldots, \varphi_{m-1}(t)\}$ is a given system of linearly independent integrable functions on $[0, r]$. The space

$$\text{Hspan } \overline{\varphi} := \{ \sum_{j=0}^{m-1} h_j(x)\varphi_j(|x|) \ : \ \{h_j\} \text{ are harmonic in } B(r)\}$$

is the harmonic span of the functions $\{\varphi_0(t), \ldots, \varphi_{m-1}(t)\}$.

Note that in the particular case $\varphi_j(t) = t^{2j}$, $j = 0, \ldots, m-1$, Hspan $\overline{\varphi}$ coincides with the space of polyharmonic functions of degree m.

For every $v \in$ Hspan $\overline{\varphi}$, an application of Lemma 1 yields,

$$\int_{B(r)} v \, dx = \sum_{j=0}^{m-1} \int_{B(r)} \varphi_j(|x|)h_j(x) \, dx = \sum_{j=0}^{m-1} a_j h_j(0) \tag{2}$$

where the coefficients a_j can be expressed in terms of univariate integration:

$$a_j = \int_{B(r)} \varphi_j(|x|) \, dx = \int_0^r \int_{S(t)} \varphi_j(|x|) d\sigma(x) = \gamma_d \int_0^r \varphi_j(t) t^{d-1} \, dt.$$

Equality (2) is a nice cubature formula. In order to make it of a certain practical value, one has to express $h_j(0)$ in terms of standard functionals of v that can be computed, like integrals of lower dimension or point evaluations of v and its derivatives. Following this approach we shall derive cubature formulae that use integrals of $\Delta^k v(x)$ or of the normal derivatives $\frac{\partial v(x)}{\partial \nu}$ over spheres $\{S(\rho_k)\}$ of distinct radii ρ_k. The cubatures have high polyharmonic and algebraic degree of precision.

2   Formulae Based on Normal Derivatives

To illustrate our method we start with a simple cubature formula that uses integrals over m distinct spheres centered at the origin.

Let $\{\rho_i\}_{i=1}^n$ be given numbers, $0 \le \rho_1 < \cdots < \rho_n \le r$. For the sake of simplicity, we shall use the abbreviated notation S_i for the sphere $S(\rho_i)$.

For a given set of functions $\overline{\varphi} = \{\varphi_0, \ldots, \varphi_{m-1}\}$, we want to construct a cubature formula of the form

$$\int_{B(r)} v(x)\, dx \approx \sum_{k=1}^{m} A_k \int_{S_k} v(x)\, d\sigma(x) \qquad (3)$$

that is exact for all functions from the Hspan $\overline{\varphi}$. In view of (2), this would be done if $h_j(0)$ can be expressed as a linear combination with constant coefficients of the integrals of v over $S_1, \ldots, S_m$. In order to establish such a link, we integrate v over S_k and get

$$\int_{S_k} v(x)\, d\sigma(x) = \sum_{i=0}^{m-1} \int_{S_k} \varphi_i(|x|) h_i(x)\, d\sigma(x) = \sum_{i=0}^{m-1} \gamma_d \varphi_i(\rho_k) \rho_k^{d-1} h_i(0), \qquad (4)$$

and this equality holds for $k = 1, \ldots, m$. Thus we arrived at a linear system of m equations in unknowns $h_0(0), \ldots, h_{m-1}(0)$. If the determinant of the system is non–zero, then it has a unique solution, and consequently, $h_i(0)$ can be expressed in terms of $\{\int_{S_k} v(x)\, d\sigma(x)\}$. It remains to insert the corresponding expressions of $h_i(0)$ in (2) and get the desired formula. Thus we have proved the following:

Proposition 1. *If the integrable functions $\varphi_0(t), \ldots, \varphi_{m-1}(t)$ and the points $\rho_1, \ldots, \rho_m$ are such that the determinant*

$$\det D(\overline{\varphi}, \overline{\rho}) := \{\varphi_i(\rho_k)\}_{i=0, k=1}^{m-1, m}$$

is distinct from zero, then there exists a cubature formula of the form (3) which is exact for all functions from Hspan $\overline{\varphi}$.

If $\det D(\overline{\varphi}, \overline{\rho}) \neq 0$, then we can get a closed form expression for the coefficients $\{A_i\}$, choosing a new, appropriate basis in span $\overline{\varphi} := \text{span } \{\varphi_0(t), \ldots, \varphi_{m-1}(t)\}$. Indeed, the condition $\det D \neq 0$ means that, for any fixed numbers $\{f_i\}$, the corresponding interpolation problem

$$\varphi(\rho_i) = f_i, \quad i = 1, \ldots, m,$$

has a unique solution $\varphi(t) \in \text{span } \overline{\varphi}$. Thus we can construct a Lagrangean basis in span φ, namely, functions $l_0(t), \ldots, l_{m-1}(t)$ defined by the interpolation conditions

$$l_i(\rho_k) = \delta_{ik}$$

(δ_{ik} being the Kronecker symbol). Then every function v from Hspan $\overline{\varphi}$ can be written in the form

$$v(x) = \sum_{i=0}^{m-1} l_i(|x|)\tilde{h}_i(x)$$

with some harmonic functions $\{\tilde{h}_i(x)\}$. Proceeding as above we arrive again at a formula of the form (2), this time with

$$a_j = \gamma_d \int_0^r l_j(t)t^{d-1}\,dt.$$

The linear system (4) reduces to

$$\int_{S_k} v(x)\,d\sigma(x) = \sum_{i=0}^{m-1} \gamma_d l_i(\rho_k)\rho_k^{d-1}\tilde{h}_i(0).$$

Because of the special choice of the basis $\{l_i\}$, the latter becomes

$$\int_{S_k} v(x)\,d\sigma(x) = \gamma_d \rho_k^{d-1}\tilde{h}_k(0)$$

and we determine $\tilde{h}_k(0)$. Consequently

$$A_k = a_k\tilde{h}_k(0) = \frac{\int_0^r l_k(t)t^{d-1}\,dt}{\rho_k^{d-1}}.$$

It is easy to notice now that the coefficients $\{A_k\}$ are actually the coefficients of an interpolatory type quadrature formula for integration of the univariate functions $f(t)$ from span $\overline{\varphi}$ over $[0,r]$. This observation was exploited in [4] and the following general statement, called a *lifting theorem* was established.

Assume that $\overline{\varphi} := \{\varphi_0(t),\ldots,\varphi_{m-1}(t)\}$ is a given system of linearly independent integrable functions on $[0,r]$. We shall say that a linear functional $L[f]$ belongs to $\mathcal{A}(B(r);\overline{\varphi})$ if

$$L[\varphi(|x|)h(x)] = h(0)L[\varphi(|x|)]$$

for each $\varphi = \varphi_0,\ldots,\varphi_{m-1}$ and every harmonic function $h(x)$ in $B(r)$.

As is a customary, span $\overline{\varphi}$ denotes the linear span of the functions $\overline{\varphi}$, that is,

$$\text{span}\,\overline{\varphi} := \{c_0\varphi_0 + \cdots + c_{m-1}\varphi_{m-1} : (c_0,\ldots,c_{m-1}) \in \mathbb{R}^m\}.$$

Set

$$L^0[\varphi] := L[\varphi(|x|)].$$

Theorem B. *Assume that $L_1, \ldots, L_N \in \mathcal{A}(B(r); \overline{\varphi})$. Then the formula*

$$\int_{B(r)} v(x) \, dx = \sum_{1}^{N} c_k L_k[v]$$

holds for every function v from Hspan $\overline{\varphi}$ *if and only if the univariate quadrature formula*

$$\gamma_d \int_0^r \varphi(t) t^{d-1} \, dt = \sum_{1}^{N} c_k L_k^0[\varphi]$$

holds for every function $\varphi \in$ span $\overline{\varphi}$.

In case $L_k[v] = \int_{S_k} v(x) \, d\sigma(x)$, by Lemma 1, we clearly have

$$L_k[\varphi(|x|)h(x)] = h(0) L_k[\varphi(|x|)]$$

and thus the result we got directly above is an immediate consequence of the lifting theorem.

Analogously, making use of Lemma 1, the following multivariate analog of the Hobby–Rice theorem (see [10]) was found in [4].

For given set of points $0 < t_1 < \cdots < t_k < 1$, let us define the *sign function* by

$$s(\overline{t}; t) := \text{sign} \ (t - t_1) \cdots (l - t_k).$$

Theorem C. *For every system $\overline{\varphi} := \{\varphi_0(t), \ldots, \varphi_{m-1}(t)\}$ of integrable functions on $[0, r]$ there exist points $0 < t_1 < \cdots < t_k < r$ with $k \leq m$ such that*

$$\int_{B(r)} s(\overline{t}; |x|) \, v(x) \, dx = 0 \quad \text{for each} \ \ v \in \text{Hspan} \ \overline{\varphi}.$$

Such an orthogonality property is related to L_1–approximation by polyharmonic functions (see [2] for a study of the harmonic case).

According to Theorem B every univariate result in the theory of numerical integration can be lifted to the multivariate case, producing cubature formulae that are exact for the corresponding harmonic span. In particular, choosing $\varphi_j(t) = t^{2j}$, one obtains formulae for polyharmonic functions. Next we shall consider multivariate formulae that are based on data consisting of integrals over spheres of the function v, its normal derivatives, or of iterated Laplacians of v. In most of the cases the study leads to difficult univariate problems that have not be considered yet. The reason is that although, for example, the value of the integral of $\Delta^k v$ over a sphere S_k is a quite natural piece of information for integration over the ball $B(r)$, the corresponding univariate data consists of values of $\Delta^k \varphi(t)$ at certain points ρ_k, which have never been used in numerical integration.

In [6] we constructed cubature formula for polyharmonic functions that was based on the evaluations

$$\left\{ \int_{S_i} \frac{\partial^k u(x)}{\partial \nu^k} \, d\sigma(x) \right\}_{k=0}^{m_i-1} , \quad i = 1, \ldots, n,$$

where $\{m_i\}$ are fixed natural numbers and $\frac{\partial}{\partial \nu}$ is the normal derivative. Next, following the approach illustrated above, we shall derive such formulae for the harmonic span of an arbitrary system of sufficiently smooth functions $\overline{\varphi}$.

The following lemma from [6] will be used.

Lemma 2. *If h is a harmonic function on $S(r)$, then*

$$\int_{S(r)} \frac{\partial^k h(x)}{\partial \nu^k} \, d\sigma(x) = 0 \quad \text{for every} \quad k = 1, 2, \ldots$$

For given functions $\overline{\varphi} = \{\varphi_0, \ldots, \varphi_{m-1}\}$, points $\overline{t} = \{t_1, \ldots, t_n\}$ and natural numbers $\overline{m} = \{m_1, \ldots, m_n\}$ with $m_1 + \cdots + m_n = m$, let us denote by $\det H(\overline{\varphi}, \overline{t}, \overline{m})$ the determinant of the matrix consisting of the rows

$$[\varphi_0^{(k)}(t_i), \varphi_1^{(k)}(t_i), \ldots, \varphi_{m-1}^{(k)}(t_i)]$$

for $i = 1, \ldots, n$ and $k = 0, \ldots, m_i - 1$. Clearly $\det H$ is the determinant corresponding to Hermite interpolation at the nodes $t_1, \ldots, t_n$ of multiplicities $m_1, \ldots, m_n$, respectively, by functions from the linear span of $\varphi_0(t), \ldots, \varphi_{m-1}(t)$.

Notice that if $\det H(\overline{\varphi}, \overline{t}, \overline{m}) \neq 0$, then the generalized polynomial $H_m(f, t)$ from span $\overline{\varphi}$, that interpolate f in the sense of Hermite, is given by the formula

$$H_m(f; t) = \sum_{i=1}^{n} \sum_{k=0}^{m_i-1} f^{(k)}(t_i) \Phi_{ik}(t)$$

where the basic functions $\{\Phi_{ik}\}$ from span $\overline{\varphi}$ are determined by the the conditions

$$\Phi_{ik}^{(l)}(t_j) = \delta_{ij}\delta_{lk}.$$

By definition (see, for example, the book of Karlin [11]), functions $\overline{\varphi} = \{\varphi_k\}_{k=0}^{m-1}$ form an *Extended Tchebycheff system* (in brief, ET system) on an interval I if and only if any generalized polynomial $c_0\varphi_0(t) + \cdots + c_{m-1}\varphi_{m-1}(t)$ has at most $m-1$ zeros in I counting the multiplicities. Equivalently, the functions $\overline{\varphi}$ form an ET system on I if and only if

$$\det H(\overline{\varphi}, \overline{t}, \overline{m}) \neq 0$$

for each $t_1 < \cdots < t_m$ in I and $m_1 + \cdots + m_n = m$.

Theorem 1. *Assume that the functions* $\varphi_0(t), \ldots, \varphi_{m-1}(t)$ *constitute an ET system on* $[0, r]$. *Then for every choice of the points* $t_1 < \cdots < t_n$ *in* $(0, r]$ *and the natural numbers* $m_1, \ldots, m_n$ *such that*

$$m_1 + \cdots + m_n = m,$$

there exists a unique cubature formula of the form

$$\int_{B(r)} v(x)\, dx \approx \sum_{i=1}^{n} \sum_{k=0}^{m_i-1} A_{ik} \int_{S_i} \frac{\partial^k}{\partial \nu^k} v(x)\, d\sigma(x) \tag{5}$$

which integrates exactly all functions v *from* Hspan $\overline{\varphi}$. *Moreover,*

$$A_{i,k} = \frac{1}{t_i^{d-1}} \int_0^r \Phi_{i,k}(t) t^{d-1}\, dt.$$

Proof. We shall derive this result from our lifting theorem (Theorem B). In order to do this, consider the functional

$$L_{ik}[v] := \int_{S_i} \frac{\partial^k}{\partial \nu^k} v(x)\, d\sigma(x).$$

For $v(x) = \varphi(|x|)h(x)$, where h is harmonic on $B(r)$ and $\varphi(t)$ is sufficiently smooth, we have

$$L_{ik}[v] = \int_{S_i} \frac{\partial^k}{\partial \nu^k} \{\varphi(|x|)h(x)\}\, d\sigma(x)$$

$$= \int_{S_i} \sum_{j=0}^{k} \binom{k}{j} \frac{\partial^{k-j}}{\partial \nu^{k-j}} \varphi(|x|) \frac{\partial^j}{\partial \nu^j} h(x) \, d\sigma(x)$$

$$= \sum_{j=0}^{k} \binom{k}{j} \int_{S_i} \frac{\partial^{k-j}}{\partial \nu^{k-j}} \varphi(|x|) \frac{\partial^j}{\partial \nu^j} h(x) \, d\sigma(x)$$

$$= \sum_{j=0}^{k} \binom{k}{j} \varphi^{k-j}(t_i) \int_{S_i} \frac{\partial^j}{\partial \nu^j} h(x) \, d\sigma(x).$$

But by the virtue of Lemma 2,

$$\int_{S_i} \frac{\partial^j}{\partial \nu^j} h(x) \, d\sigma(x) = 0$$

for $j > 0$. Therefore

$$L_{ik}[\varphi(|x|)h(x)] = \varphi^{(k)}(t_i) \int_{S_i} h(x) \, d\sigma(x) = \int_{S_i} \varphi^{(k)}(t) \, d\sigma(x) h(0) = L_{ik}^0[\varphi] \, h(0).$$

Thus the functionals L_{ik} satisfy the requirement of Theorem B to belong to $\mathcal{A}(B(r); \overline{\varphi})$. Then the existence and uniqueness of the cubature formula (5) is related to the existence and uniqueness of the univariate quadrature formula

$$\gamma_d \int_0^r \varphi(t) t^{d-1} \, dt = \sum_{i=1}^{n} \sum_{k=0}^{m_i-1} A_{ik} \int_{S_i} \varphi^{(k)}(t) \, d\sigma(x)$$

which reduces to

$$\int_0^r \varphi(t) t^{d-1} \, dt = \sum_{i=1}^{n} \sum_{k=0}^{m_i-1} A_{ik} t_i^{d-1} \varphi^{(k)}(t_i).$$

But the latter is an Hermitian type quadrature formula in span $\overline{\varphi}$ and as is well-known it exists and is unique if and only if $\det D(\overline{\varphi}, \overline{t}, \overline{m}) \neq 0$. Besides, the coefficients $A_{ik} t_i^{d-1}$ are obtained as integrals of the basic functions $\Phi_{ik}(t)$. The proof is completed.   □

The polyharmonic case studied in [6] can be derived from the above theorem. To do this we need the following simple lemma (see [11, p. 18, Theorem 2.1]).

Lemma 3. *If the functions $\varphi_0(t), \ldots, \varphi_{m-1}(t)$ constitute an ET system on an interval I and the function $\theta : J \to I$ satisfies $\theta'(\tau) > 0$ on J, then the functions $\varphi_0(\theta(\tau)), \ldots, \varphi_{m-1}(\theta(\tau))$ form an ET system on J.*

Now, choose $\varphi_j(t) = t^{2j}$, $j = 0, \ldots, m-1$. Then the harmonic span of $\varphi_0, \ldots, \varphi_{m-1}$ coincides with the space $H_m(B)$ of polyharmonic functions on $B(r)$ of degree m. Besides, the functions $\{t^k\}$ constitute an ET system on $[0,1]$ and the function $\theta(\tau) := t^2$ has a positive derivative on $(0,1]$. Hence, by Lemma 3, the functions $\varphi_0(t^2), \ldots, \varphi_{m-1}(t^2)$ form an ET system on $(0,1]$ and therefore $\det H(\overline{\varphi}, \overline{t}, \overline{m}) \neq 0$ for any $\overline{m}$ and points $0 < t_1 < \cdots < t_n \leq 1$. Then we derive from Theorem 2 the existence and uniqueness of a cubature formula of form (5) which is exact for all polyharmonic functions of degree m.

To find the coefficients A_{ik} of the formula in this particular case we construct the algebraic polynomial $Q_{ik}(t)$ of degree $2m-1$ that satisfies the interpolation conditions

$$Q_{ik}^{(j)}(t_l) = \delta_{il}\delta_{kj}, \quad Q_{ik}^{(j)}(-t_l) = (-1)^j \delta_{il}\delta_{kj}, \quad l = 1, \ldots, n.$$

Clearly Q_{ik} is determined uniquely, it is even, and thus of degree $2m-2$. Then it can be written as $Q_{ik}(t) = P_{ik}(t^2)$ with a certain polynomial P_{ik} of degree $m-1$. Thus $Q_{ik}(t)$ belongs to span $\overline{\varphi}$ and satisfies the conditions

$$Q_{ik}^{(j)}(t_l) = \delta_{il}\delta_{kj}, \quad l = 1, \ldots, n, \; j = 0, \ldots, m_l - 1.$$

Therefore, in the polyharmonic case we have

$$A_{ik} = \frac{1}{t_i^{d-1}} \int_0^1 Q_{ik}(t) t^{d-1} \, dt,$$

which is the formula obtained in [6].

3 Formulae Based on Iterated Laplacians

In this section we shall consider the question of existence of a cubature formula of the form

$$\int_{B(r)} v(x) \, dx \approx \sum_{i=1}^{n} \sum_{k=0}^{m_i - 1} B_{ik} \int_{S(\rho_i)} \Delta^k v(x) \, d\sigma(x),$$

involving Laplace operators of consecutive orders with fixed $\{\rho_i\}$ and $m := m_1 + \cdots + m_n$ free coefficients, which integrates exactly all functions from the harmonic span of a system of m given functions.

Let us recall first two auxiliary results from [4] concerning the differential operator Δ^k.

Lemma 4. *Let $\varphi \in C^{2p}[0, r^2]$. Then for each $x \in B(r)$ we have*

$$\Delta^p \varphi(|x|^2) = \sum_{j=0}^{p} c_{pj} |x|^{2j} \varphi^{(p+j)}(|x|^2)$$

with certain positive constants $\{c_{pj}\}$ which do not depend on φ.

Lemma 5. *For a fixed $p \geq 0$, let φ be any function from $C^{2p}[0, r^2]$. Then, for every harmonic function h in $B(r)$, we have*

$$\Delta^p[\varphi(|x|^2)h(x)] = \Delta^p[\varphi(|x|^2)]\, h(x) + \sum_{j=1}^{2^p-1} \alpha_{pj}(|x|^2)g_{pj}(h; x)$$

with certain functions $\{\alpha_{pj}(\tau)\}$ which do not depend on h, and harmonic functions $g_{pj}(h; x)$ satisfying the condition

$$g_{pj}(h; 0) = 0, \quad j = 0, \ldots, 2^p - 1.$$

For each sufficiently smooth real function f and a real number $\rho \geq 0$ we set

$$\Delta f(\rho^2) := \Delta f(t^2)\big|_{t=\rho} := \Delta f(|x|^2)\big|_{|x|=\rho}.$$

As seen from Lemma 4, the last expression takes the same value for every x satisfying $|x| = \rho$. Thus the notation $\Delta f(\rho^2)$ is well defined. As a consequence of this definition we get the rule

$$\Delta f(t^2) = 4t^2 f''(t^2) + 2df'(t^2)$$

which is a known represntation of the radial part of Δ.

Assume that $m_1, \ldots, m_n$ are fixed natural numbers and $m = m_1 + \cdots + m_n$. With any given system of sufficiently smooth functions $\overline{\varphi} := \{\varphi_0(\tau), \ldots, \varphi_{m-1}(\tau)\}$ on $[0, r]$ and numbers $0 \leq \rho_1 < \cdots < \rho_n \leq r$ we associate the matrix

$$V(\overline{\varphi}; \rho_1, \ldots, \rho_n) := \begin{bmatrix} V_1 \\ V_2 \\ \vdots \\ V_n \end{bmatrix}$$

where the blocks V_i are of the form

$$V_i := \begin{bmatrix} \varphi_0(\rho_i^2) & \varphi_1(\rho_i^2) & \cdots & \varphi_{m-1}(\rho_i^2) \\ \Delta\varphi_0(\rho_i^2) & \Delta\varphi_1(\rho_i^2) & \cdots & \Delta\varphi_{m-1}(\rho_i^2) \\ \cdots & \cdots & \cdots & \cdots \\ \Delta^{m_i-1}\varphi_0(\rho_i^2) & \Delta^{m_i-1}\varphi_1(\rho_i^2) & \cdots & \Delta^{m_i-1}\varphi_{m-1}(\rho_i^2) \end{bmatrix}$$

Our existence theorem follows.

Theorem 2. *Let* $\overline{\varphi} := \{\varphi_i(t^2)\}_{i=0}^{m-1}$ *be any system of functions from* $C^{2m-2}[0,r]$ *satisfying the condition*

$$\det V(\overline{\varphi}; \rho_1, \ldots, \rho_n) \neq 0.$$

Then there exist coefficients $\{a_{qp}(\overline{\rho})\}$ *such that the cubature formula*

$$\int_{B(r)} v(x)\, dx \approx \sum_{q=1}^{n} \sum_{p=0}^{m_q-1} a_{pq}(\overline{\rho}) \int_{S_q} \Delta^p v(x)\, d\sigma(x)$$

integrates exactly all functions from Hspan $\{\varphi_i(t^2)\}$. *The coefficients* $\{a_{qp}(\overline{\rho})\}$ *are determined as solutions of a linear system.*

Proof. The result can be derived from the lifting theorem. We prefer to give a short direct proof.

Let v be any function from the class Hspan $\{\varphi_i(t^2)\}$. By definition, v can be represented in the form

$$v(x) = \sum_{i=0}^{m-1} \varphi_i(|x|^2) h_i(x) \tag{6}$$

with some harmonic functions $h_i(x)$ on $B(r)$. Integrating the both sides of the last equality over $B(r)$, we get (2). Our next task is to express the quantities $\{h_i(0)\}$ in terms of the integrals

$$\left\{ \int_{S_q} \Delta^p v(\xi)\, d\sigma(\xi) \right\}_{q=1,\, p=0}^{n \quad m_q-1}.$$

In order to do this, we apply the operator Δ^p to the both sides of (6) and integrate over S_q:

$$\int_{S_q} \Delta^p v(x)\, dx = \sum_{i=0}^{m-1} \int_{S_q} \Delta^p[\varphi_i(|x|^2)\, h_i(x)]\, d\sigma(x). \tag{7}$$

By Lemma 5, the corresponding functions $g_{pj}(h_i; x)$ in the representation of $\Delta^p[\varphi_i(|x|^2) h_i(x)]$ are harmonic and vanish at zero. Consequently, the integrals over S_q of g_{pj} equal zero. Thus,

$$\int_{S_q} \Delta^p[\varphi_i(|x|^2)\, h_i(x)]\, d\sigma(x) = \int_{S_q} \Delta^p[\varphi_i(|x|^2)]\, h_i(x)\, d\sigma(x)$$

$$+ \sum_{j=1}^{2^p-1} \alpha_{pj}(\rho_i^2) \int_{S_q} g_{pj}(h_i; x)\, d\sigma(x).$$

$$= \Delta^p[\varphi_i(t^2)]\Big|_{t=\rho_q} \int_{S_q} h_i(x)\, d\sigma(x)$$

$$= \Delta^p[\varphi_i(t^2)]\Big|_{t=\rho_q} \gamma_d \rho_q^{d-1} h_i(0).$$

Next we substitute the last expression in (7) and arrive at the equation

$$\int_{S_q} \Delta^p v(x)\, d\sigma(x) = \gamma_d \rho_q^{d-1} \sum_{i=0}^{m-1} \Delta^p[\varphi_i(t^2)]\Big|_{t=\rho_q} h_i(0).$$

Since it holds for every $q = 1, \ldots, n$ and $p = 0, \ldots, m_q - 1$, we get a system of linear equations with respect to $\{h_i(0)\}_{i=0}^{m-1}$. By assumption, the matrix $V(\overline{\varphi}; \rho_1, \ldots, \rho_n)$ of this system is not singular. Thus the quantities $\{h_i(0)\}$ can be expressed in terms of $\{\int_{S_q} \Delta^p v(x)\, d\sigma(x)\}$. Then (2) results in a cubature of the desired form. The proof is complete.   □

The condition $\det V(\overline{\varphi}; \rho_1, \ldots, \rho_n) \neq 0$ is equivalent to the requirement for regularity of the interpolation problem

$$\Delta^p g(t)\Big|_{t=\rho_q} = f_{pq}, \quad q = 1, \ldots, n, \; p = 0, \ldots, m_q - 1,$$

by functions g from the harmonic span of $\{\varphi_0(t^2), \ldots, \varphi_{m-1}(t^2)\}$. As far as we know this problem was not studied in the literature. We were not able to characterized those systems $\overline{\varphi}$ for which the corresponding interpolation problem is regular for each choice of the points $\rho_1 < \cdots < \rho_n$ in $(0, r]$. Even in the polyharmonic case, when $\varphi_j(\tau) = \tau^j$, $j = 0, \ldots m - 1$, and the question is reduced to a polynomial interpolation problem, the regularity of $V(\overline{\varphi}; \rho_1, \ldots, \rho_n)$ is still not settled. In the case $n = 1$ it is seen with the help of Lemma 4 that the determinant

$$\det V(\overline{\varphi}; \rho) := \begin{bmatrix} \varphi_0(\rho^2) & \varphi_1(\rho^2) & \cdots & \varphi_{m-1}(\rho^2) \\ \Delta\varphi_0(\rho^2) & \Delta\varphi_1(\rho^2) & \cdots & \Delta\varphi_{m-1}(\rho^2) \\ \cdots & \cdots & \cdots & \cdots \\ \Delta^{m-1}\varphi_0(\rho^2) & \Delta^{m-1}\varphi_1(\rho_i^2) & \cdots & \Delta^{m-1}\varphi_{m-1}(\rho^2) \end{bmatrix}$$

is triangular, with non–zero elements on the diagonal, and thus it is distinct from zero. Then, as a corollary of Theorem 2, we get the following result obtained in [4].

Corollary 1. *Let ρ be any point from $(0, r]$. Then there exist coefficients $\{a_k(\rho)\}_0^{m-1}$ such that the cubature formula*

$$\int_{B(r)} u(x)\, dx = \sum_{k=0}^{m-1} a_k(\rho) \frac{1}{\gamma_n \rho^{n-1}} \int_{S(\rho)} \Delta^k u(x)\, d\sigma(x) \tag{8}$$

holds for every polyharmonic function u of degree m. The coefficients $\{a_k(\rho)\}$ are determined as solutions of the linear system

$$a_0(\rho)\rho^{2j} + a_1(\rho)\Delta\rho^{2j} + \cdots + a_j(\rho)\,\Delta^j \rho^{2j} = b_j, \quad j = 0, \ldots, m-1,$$

with

$$b_j := \gamma_d \int_0^r t^{2j+d-1}\, dt = \gamma_d \frac{r^{d+2j}}{d+2j}, \quad j = 0, \ldots, m-1.$$

Let us mention another example in which a cubature formula for polyharmonic functions can be constructed.

Theorem 3. *Assume that $\{m_i\}$, $m_1 + \cdots + m_n = m$, $(m_0 := 0)$. Then, for each choice of the points $0 < \rho_1 < \cdots < \rho_n \leq r$, there exist coefficients $\{b_{ik}\}$ such that the cubature formula*

$$\int_{B(r)} u(x)\, dx = \sum_{i=1}^n \sum_{k=m_{i-1}}^{m_i-1} b_{ik}(\rho) \int_{S(\rho_i)} \Delta^k u(x)\, d\sigma(x)$$

holds for every polyharmonic function u of degree m.

Proof. We only sketch the proof, following that of the previous theorem and keeping in mind that $\varphi_j(\tau) = \tau^j$, $j = 0, \ldots m-1$. In this case one arrives at a linear system with a determinant consisting of the blocks

$$V_i := \begin{bmatrix} \Delta^{m_i-1}\varphi_0(\rho_i^2) & \Delta^{m_i-1}\varphi_1(\rho_i^2) & \cdots & \Delta^{m_i-1}\varphi_{m-1}(\rho_i^2) \\ \Delta^{m_i-1+1}\varphi_0(\rho_i^2) & \Delta^{m_i-1+1}\varphi_1(\rho_i^2) & \cdots & \Delta^{m_i-1+1}\varphi_{m-1}(\rho_i^2) \\ \cdots & \cdots & \cdots & \cdots \\ \Delta^{m_i-1}\varphi_0(\rho_i^2) & \Delta^{m_i-1}\varphi_1(\rho_i^2) & \cdots & \Delta^{m_i-1}\varphi_{m-1}(\rho_i^2) \end{bmatrix}.$$

By Lemma 4,

$$\Delta^p \varphi_i(|x|^2) = \sum_{j=0}^p c_{pj} |x|^{2j} \varphi_i^{(p+j)}(|x|^2)$$

and since in the polyharmonic case $\varphi_i(\tau) = \tau^i$, $i = 0, \ldots, m-1$, we see that

$$\varphi_i^{(p+j)}(|x|^2) = 0 \quad \text{for} \quad p > i,$$

$$\Delta^p \varphi_p(|x|^2) = \varphi_p^{(p)}(|x|^2) = c_{p0}|x|^{2p}p! > 0.$$

Therefore the determinant is triangular, and consequently non–zero. The proof is complete. □

Corollary 1 follows also from Theorem 3 setting $\rho_1 = \cdots = \rho_n$. Further, letting ρ to approach zero in (8), we can derive from Corollary 1 a cubature formula based on the values $\{\Delta^j u(0)\}_{j=0}^{m-1}$. More precisely, we get the famous Pizzetti formula [15]:

$$\int_{B(r)} u(x)\, dx = \pi^{d/2}\, r^d \sum_{k=0}^{m-1} \frac{r^{2k}}{2^{2k}\Gamma(d/2+k+1)} \frac{\Delta^k u(0)}{k!}$$

which holds for every polyharmonic function u of degree m (see also [14] or [3] for another proof).

The result of Pizzetti can be extended to harmonic spans of wide class of smooth functions. This was done recently in [4]. We give the result below, for completeness.

Denote by $W[\overline{\varphi}; t]$ the Wronskian of the system $\overline{\varphi}$ at t, that is,

$$W[\overline{\varphi}; t] := \begin{bmatrix} \varphi_0(t) & \varphi_0'(t) & \cdots & \varphi_0^{(m-1)}(t) \\ \varphi_1(t) & \varphi_1'(t) & \cdots & \varphi_1^{(m-1)}(t) \\ \cdots & \cdots & \cdots & \cdots \\ \varphi_{m-1}(t) & \varphi_{m-1}'(t) & \cdots & \varphi_{m-1}^{(m-1)}(t) \end{bmatrix}.$$

Theorem 4. *Let $\overline{\varphi} := \{\varphi_0, \ldots, \varphi_{m-1}\}$ be any system of functions from $C^{2m-2}[0, r^2]$ satisfying the condition $W[\overline{\varphi}; 0] \neq 0$. Then there exist coefficients $\{A_k\}_0^{m-1}$ such that the formula*

$$\int_{B(r)} v(x)\, dx = \sum_{k=0}^{m-1} A_k \Delta^k v(0)$$

holds for every $v \in \operatorname{Hspan}\{\varphi_0(t^2), \ldots, \varphi_{m-1}(t^2)\}$. The coefficients are determined as solutions of a linear system.

Proof. By the lifting theorem, the cubature formula holds if and only if

$$\gamma_d \int_0^r \varphi_j(t^2)\, t^{d-1}\, dt = \sum_{k=0}^{m-1} A_k \Delta^k [\varphi_j(t^2)]\Big|_{t=0}$$

for each $j = 0, \ldots, m-1$. By Lemma 4,

$$\Delta^k [\varphi_j(t^2)]\Big|_{t=0} = c_{k0}\, \varphi_j^{(k)}(0).$$

Thus, one can find the coefficients $\{A_k\}$ from the linear system

$$\gamma_d \int_0^r \varphi_j(t^2)\, t^{d-1}\, dt = \sum_{k=0}^{m-1} A_k\, c_{k0}\, \varphi_j^{(k)}(0), \quad j = 0, \ldots, m-1,$$

which, in view of the assumption, has a non–zero determinant. The proof is complete. $\square$

In the polyharmonic case, when $\varphi_k(\tau) = \tau^k$, $k = 0, \ldots, m-1$, the system is a diagonal one and hence $\{A_k\}$ can be found explicitly. Indeed, using the formula

$$c_{k0} = 2^{2k}(d/2)_k = 2^{2k}\frac{\Gamma(d/2+k)}{\Gamma(d/2)}$$

(found in [4]) and taking into account that $\varphi_j^{(k)}(0) = k!\, \delta_{kj}$, we obtain

$$A_k = \frac{\gamma_d \int_0^r t^{2k+d-1}\, dt}{c_{k0}\, k!} = \frac{\pi^{d/2}\, r^{2k}}{2^{2k}\, \Gamma(d/2+k+1)} \cdot \frac{r^d}{k!},$$

which is the familiar formula for the Pizzetti coefficients.

4 Formulae of Mixed Type

Recently Dimitrov [7] constructed a cubature formula on the unit ball $B := B(1)$ $(S := S(1))$ of the form

$$\int_B u(x)\, dx \approx \sum_{j=0}^{[(m-1)/2]} C_j^{(e)} \int_S \Delta^j u(x)\, d\sigma(x)$$

$$+ \sum_{j=0}^{[m/2]-1} C_j^{(o)} \int_S \frac{\partial}{\partial \nu} \Delta^j u(x)\, d\sigma(x),$$

which integrates exactly all polyharmonic functions of degree m. The information data here is mixed, values of the Laplacians and the normal derivatives are used. That is why we refer to such formulae as "formulae of mixed type". Dimitrov's formula is based on integrals over the boundary S of the unit ball. The method demonstrated in the previous sections allows us to study formulae using integrals over n distinct spheres that are exact for the harmonic span of any ET system.

We shall prove the following.

Theorem 5. *Assume that the functions $\varphi_0(t), \ldots, \varphi_{m-1}(t)$ constitute an ET system on $[0, r]$. Then for every choice of the points $\rho_1 < \cdots < \rho_n$ in $(0, r]$ and the integers $m_1, \ldots, m_n$, $m_1 + \cdots + m_n = m$, there exists a unique cubature formula of the form*

$$\int_{B(r)} v(x)\, dx \approx \sum_{i=1}^{n} \sum_{k=1}^{m_i-1} C_{ik}^{(e)} \int_{S_i} \Delta^k v(x)\, d\sigma(x) \tag{9}$$

$$+ \sum_{i=1}^{n} \sum_{k=0}^{m_i-1} C_{ik}^{(o)} \int_{S_i} \frac{\partial}{\partial \nu} \Delta^k v(x)\, d\sigma(x)$$

which integrates exactly all functions v from the harmonic span of $\{\varphi_k(t^2)\}_{k=0}^{2m-1}$. The coefficients $\{C_{ik}^{(e)}, C_{ik}^{(o)}\}$ can be computed as solutions of a linear system.

Proof. To apply the lifting theorem we need first to check that the functional

$$L[v] := \int_{S_i} \frac{\partial}{\partial \nu} \Delta^p v(x)\, d\sigma(x)$$

belongs to $\mathcal{A}(B(r); \overline{\varphi})$. Indeed, for $v(x) = \varphi(|x|^2)h(x)$, we have on the basis of Lemma 5

$$L[\varphi(|x|^2)h(x)]$$

$$= \int_{S_i} \frac{\partial}{\partial \nu} \left\{ \Delta^p[\varphi(|x|^2)]h(x) + \sum_{j=1}^{2^p-1} \alpha_{pj}(|x|^2) g_{pj}(h; x) \right\} d\sigma(x)$$

$$= \int_{S_i} \left\{ \frac{\partial}{\partial \nu} \Delta^p[\varphi(|x|^2)] \cdot h(x) + \Delta^p[\varphi(|x|^2)] \frac{\partial}{\partial \nu} h(x) \right\} d\sigma(x)$$

$$+ \int_{S_i} \sum_{j=1}^{2^p-1} \left\{ \frac{\partial}{\partial \nu} \{\alpha_{pj}(|x|^2)\} g_{pj}(h; x) + \alpha_{pj}(|x|^2) \frac{\partial}{\partial \nu} g_{pj}(h; x) \right\} d\sigma(x).$$

Now using the properties of the functions g_{pj}, described in Lemma 5, and applying Lemma 2, we get

$$L[\varphi(|x|^2)h(x)] = L^0[\varphi(t^2)]h(0).$$

Thus $L \in \mathcal{A}(B(r); \overline{\varphi})$. Then, by Theorem B, the problem is reduced to the existence of a quadrature formula of the form

$$
\gamma_d \int_0^r \varphi_q(t^2)t^{d-1}\,dt = \sum_{i=1}^n \sum_{k=0}^{m_i-1} C_{ik}^{(o)} \int_{S_i} \frac{\partial}{\partial \nu} \Delta^k \varphi_q(t^2)\,d\sigma(x)
$$
$$
+ \sum_{i=1}^n \sum_{k=1}^{m_i-1} C_{ik}^{(e)} \int_{S_i} \Delta^k[\varphi_i(|x|^2)]\,d\sigma(x)
$$

for $q = 0,\ldots,2m-1$. After an obvious simplification it becomes

$$
\int_0^r \varphi_q(t^2)t^{d-1}\,dt = \sum_{i=1}^n \sum_{k=0}^{m_i-1} C_{ik}^{(o)} \rho_i^{d-1} \frac{d}{dt}\Delta^k \varphi_q(t^2)\Big|_{t=\rho_i}
$$
$$
+ \sum_{i=1}^n \sum_{k=1}^{m_i-1} C_{ik}^{(e)} \rho_i^{d-1}\Delta^k[\varphi_q(t^2)]\Big|_{t=\rho_i}.
$$

This is a linear system with respect to the coefficients $\{C_{ik}\}$. The determinant of this system can be written in the form $[M_1,\ldots,M_n]^T$ with blocks

$$
M_i := \begin{bmatrix}
\varphi_0(t^2) & \varphi_1(t^2) & \cdots & \varphi_{m-1}(t^2) \\
\frac{d}{dt}\varphi_0(t^2) & \frac{d}{dt}\varphi_1(t^2) & \cdots & \frac{d}{dt}\varphi_{m-1}(t^2) \\
\Delta\varphi_0(t^2) & \Delta\varphi_1(t^2) & \cdots & \Delta\varphi_{m-1}(t^2) \\
\frac{d}{dt}\Delta\varphi_0(t^2) & \frac{d}{dt}\Delta\varphi_1(t^2) & \cdots & \frac{d}{dt}\Delta\varphi_{m-1}(t^2) \\
\Delta^2\varphi_0(t^2) & \Delta^2\varphi_1(t^2) & \cdots & \Delta^2\varphi_{m-1}(t^2) \\
\vdots & \vdots & \cdots & \vdots \\
\frac{d}{dt}\Delta^{m_i-1}\varphi_0(t^2) & \frac{d}{dt}\Delta^{m_i-1}\varphi_1(t^2) & \cdots & \frac{d}{dt}\Delta^{m_i-1}\varphi_{m-1}(t^2)
\end{bmatrix}
$$

at $t = \rho_i$. Next, taking into account the differentiation formula

$$\Delta\varphi(t^2) = 4t^2\varphi''(t^2) + 2d\varphi'(t^2),$$

we see that the element of index q in the consecutive rows of the block M_i is a linear combination of the following derivatives, respectively,

$$\varphi_q(t^2)$$
$$\frac{d}{dt}\varphi_q(t^2)$$
$$\frac{d}{dt}\varphi_q(t^2), \frac{d^2}{dt^2}\varphi_q(t^2)$$
$$\frac{d}{dt}\varphi_q(t^2), \frac{d^2}{dt^2}\varphi_q(t^2), \frac{d^3}{dt^3}\varphi_q(t^2)$$
$$\frac{d}{dt}\varphi_q(t^2), \frac{d^2}{dt^2}\varphi_q(t^2), \ldots, \frac{d^4}{dt^4}\varphi_q(t^2)$$
$$\cdots \qquad \cdots \qquad \cdots$$
$$\frac{d}{dt}\varphi_q(t^2), \frac{d^2}{dt^2}\varphi_q(t^2), \ldots, \frac{d^{2m_i-1}}{dt^{2m_i-1}}\varphi_q(t^2)$$

Using elementary property of determinants one can show that this portion of the determinant can be transformed to the one corresponding to Hermite interpolation problem of multiplicity $2m_i$ at the point ρ_i. Thus the problem of constructing a cubature formula of the form (9) is equivalent to the construction of a cubature based on integrals of consecutive normal derivatives up to order $2m_i - 1$ over a sphere of radius ρ_i, respectively. But we showed already in Theorem 1 that such a cubature can be constructed uniquely. The theorem is proved. □

5　Gaussian Type Cubature Formulae

In the previous section we constructed cubature formulae that use N pieces of information (integrals over spheres of radius $\rho_1 < \cdots < \rho_n$, respectively, and integrate exactly all functions from a space of "harmonic" dimension N, that is, all functions of the form

$$\sum_{i=1}^{N} \varphi_i(|x|)h_i(x)$$

with harmonic $h_i(x)$. In particular, we obtained formulae of preassigned type of polyharmonic degree of precision (in brief, PDP) N. Now we shall consider the Gaussian extremal problem:

For what choice of the radii $\{\rho_i\}$ does the corresponding cubature formula attain its maximal polyharmonic degree of precision?

Our first example is related to the classical Gauss formula. As is well–known Gauss [8] proved that the interpolatory quadrature formula

$$\int_{-1}^{1} f(t)\, dt \approx \sum_{k=1}^{m} a_k f(\tau_k)$$

at the zeros $\{\tau_k\}_{k=1}^{m}$ of the Legendre orthogonal polynomial of degree m has algebraic degree of precision $2m - 1$ and this is the highest degree that can be achieved by a formula based on m point evaluations. A remarkable extension of the Gauss result was given by Krein [12], who proved the same for any Tchebycheff system of continuous functions $\varphi_0(t), \ldots, \varphi_{2m-1}(t)$ on $[-1, 1]$.

Lifting the Krein result with the help of Theorem B, the following multivariate analog of the Gauss formula was obtained in [4]. For the sake of simplicity we restrict ourselves to integration over the unit ball $B := B(1)$.

Proposition 2. *Let* $\overline{\varphi} := \{\varphi_0(t), \ldots, \varphi_{2m-1}(t)\}$ *be a Tchebycheff system of continuous functions on* $[0, 1]$. *Then there exists a unique cubature formula of the form*

$$\int_{B} v(x)\, dt \approx \sum_{k=1}^{m} A_k \int_{S(t_k)} v(x)\, d\sigma(x)$$

which integrates exactly all functions $v \in \mathrm{Hspan}\ \overline{\varphi}$.

Proof. By Theorem B the problem is equivalent to the existence and uniqueness of a quadrature formula

$$\int_{0}^{1} \varphi(t) t^{d-1}\, dt \approx \sum_{k=1}^{m} A_k t_k^{d-1} \varphi_k(t_k) \tag{10}$$

that integrates exactly all functions $\varphi = \varphi_0, \varphi_1, \ldots \varphi_{2m-1}$ on $[0, 1]$, and therefore to the Krein result. $\square$

In particular, choosing

$$\varphi_j(t) = t^{2k}, \quad j = 0, 1, \ldots, 2m - 1,$$

we get a formula of highest PDP, which was first established in [6].

Proposition 3. *The cubature formula*

$$\int_{B} u(x)\, dt \approx \sum_{k=1}^{m} G_k \int_{S(\rho_k)} u(x)\, d\sigma(x)$$

with nodes $\{\rho_k\}_{k=1}^m$ at the positive zeros of the orthogonal polynomial P_{2m} on $[-1,1]$ with weight $|t|^{d-1}$ and coefficients

$$G_k = \frac{1}{t_k^{d-1}(P'_{2m}(t_k))^2} \int_{-1}^1 \left(\frac{P_{2m}}{t - t_k}\right)^2 dt$$

integrates exactly all polyharmonic functions u of degree $2m$. Moreover, this is the unique cubature formula of this type with $PDP = 2m$.

Proof. By the lifting theorem the problem is reduced to the characterization of quadrature formulae of the form

$$\int_0^1 t^{2j} t^{d-1} \, dt \approx \sum_{k=1}^m G_k t_k^{d-1} t_k^{2j}$$

for maximal number of consecutive even powers t^{2j}. Consider the equivalent quadrature on $[-1,1]$

$$\int_{-1}^1 f(t)|t|^{d-1} \, dt \approx \sum_{k=1}^m G_k t_k^{d-1} f(t_k) + \sum_{k=1}^m G_k t_k^{d-1} f(-t_k) \qquad (11)$$

for even polynomials f. Clearly, because of the symmetry, the last formula is exact for all odd polynomials. Thus, if (11) is exact for all even polynomials of degree $2N$, it would have algebraic degree of precision (in brief, ADP) equal to $2N + 1$. On the other side, the Gauss quadrature formula with $2m$ nodes has ADP $= 4m - 1$. Moreover, the Gauss nodes are symmetric and hence the Gauss formula is of the form (11). Therefore the Gauss formula with weight t^{d-1} on $[-1,1]$ supplies the highest ADP for (11), equal to $4m - 1$ and this is the only formula with this extremal property. Consequently, the corresponding formula (10) integrates exactly all even polynomials of degree $4m - 2$ and by the lifting theorem this induces a formula of the form described in Proposition 3 of highest $PDP = 2m$.

Let P_{2m} be the Legendre polynomial of degree $2m$ that is orthogonal on $[-1,1]$ with a weight $|t|^{d-1}$. Let $-\rho_m < \cdots < -\rho_1 < \rho_1 < \cdots < \rho_m$ be the zeros of P_{2m}. It remains to observe that the coefficients of the Gauss quadrature (11) are given by the formula

$$t_q^{d-1} G_k = \frac{1}{(P'_{2m}(\rho_k))^2} \int_{-1}^1 \left(\frac{P_{2m}(t)}{t - \rho_k}\right)^2 dt.$$

The proposition is proved.   □

Since every algebraic polynomial of degree $2m - 1$ is a polyharmonic function of degree m, we conclude that the formula in Proposition 3 integrates exactly all polynomials of degree $4m - 1$. This fact was established by Kantarovich and Lusternik (see, for example, [13]).

The next more general result was given recently in [6].

Theorem 6. *For any given set of odd multiplicities $m_1, \ldots, m_n$ with $m = m_1 + \cdots + m_n$ there exists a unique cubature formula of the form*

$$\int_B v(x)\, dx \approx \sum_{i=1}^{n} \sum_{k=0}^{m_i-1} A_{ik} \int_{S(R_i)} \frac{\partial^k}{\partial \nu^k} v(x)\, d\sigma(x)$$

that is exact for all polyharmonic functions of degree $(m + n)$. The nodes $R_1, \ldots R_n$ of this cubature are located at the positive zeros of the polynomial

$$\Omega(t^2) = \prod_{k=1}^{n} (t^2 - t_k^2)^{m_i}$$

satisfying the orthogonality relations:

$$\int_{-1}^{1} \Omega(t^2) Q(t^2) |t|^{d-1}\, dt = 0 \quad \text{for every} \quad Q \in \pi_{m-1}.$$

We shall give here another proof based on the lifting theorem.

By Theorem B the problem is reduced to a quadrature of the form

$$\int_0^1 f(t) t^{d-1}\, dt \approx \sum_{i=1}^{n} \sum_{k=0}^{m_i-1} a_{ik} f^{(k)}(t_k) \tag{12}$$

of highest ADP, say N. The existence of such a cubature would imply the existence of the corresponding cubature on $[-1, 0]$, namely

$$\int_{-1}^{0} f(t)(-t)^{d-1}\, dt \approx \sum_{i=1}^{n} \sum_{k=0}^{m_i-1} (-1)^k a_{ik} f^{(k)}(-t_k)$$

with the same ADP. Combining both of them we would get a cubature of the form

$$\int_{-1}^{1} f(t) |t|^{d-1}\, dt \approx \sum_{i=1}^{n} \sum_{k=0}^{m_i-1} a_{ik} f^{(k)}(t_k) + \sum_{i=1}^{n} \sum_{k=0}^{m_i-1} (-1)^k a_{ik} f^{(k)}(-t_k)$$

of ADP $= N$. According to a result due to Ghizzetti and Ossicini [9] such a quadrature exists, it is unique, and has ADP $= 2m + 2n - 1$. Because of the uniqueness its nodes are symmetric and thus, it is of the form above. Then the parameters of (12) are determined. As is known from [9] the nodes t_k are the unique solution of the minimization problem

$$\int_{-1}^{1} |t|^{d-1} \prod_{k=1}^{n} (t^2 - t_k^2)^{m_i+1} \, dt \rightarrow \min \text{ over } 0 < t_1 < \cdots < t_n \leq 1.$$

The solution is uniquely characterized by the orthogonal property

$$\int_{-1}^{1} |t|^{d-1} \prod_{k=1}^{n} (t^2 - t_k^2)^{m_i} P(t) \, dt = 0$$

for all polynomials P of degree $2n - 1$. Since the orthogonality to all odd powers t^{2j-1} is quaranteed by the symmetricity of the formula, it is enough to require orthogonality to polynomials of the form $P(t) = Q(t^2)$, $Q \in \pi_{m-1}$, as stated in the theorem.

The coefficients $\{A_{ik}\}$ of the extremal cubature are related to the coefficients $\{a_{ik}\}$ of the Ghizzetti–Ossicini formula by $A_{ik} R_i^{d-1} = a_{ik}$.   □

Note that an extension of the last result to integration of harmonic span of ET systems can be derived by lifting the corresponding univariate result from [5] concerning generalized Gaussian quadrature formulae with multiple nodes.

Finally, we present an open question related to the cubature formula (8) that was constructed in Corollary 1. For any given $\rho \in (0,]$, the coefficients $a_k(\rho)$, $k = 0, \ldots, m - 1$, can be found solving a triangular system of m equations. Consider the last coefficient $a_{m-1}(\rho)$ as a function of ρ. *Is there a particular* $\rho = \rho_{m-1}^*$ *for which the coefficient* $a_{m-1}(\rho)$ *vanishes? Is there a unique such* ρ? If so, the corresponding cubature (8) would use only $m - 1$ (instead of m) integrals over the sphere of radius ρ_{m-1}^* and would be exact for all polyharmonic functions of degree m. Thus the resulting cubature would be Gaussian. The question was considered for small m in [4]. We found there that

$$a_0(\rho) = \gamma_d \frac{r^d}{d}$$

$$a_1(\rho) = \gamma_d \frac{r^d}{2d} \left(\frac{r^2}{d+2} - \frac{\rho^2}{d} \right)$$

$$a_2(\rho) = \gamma_d \frac{r^d}{8d(d+2)} \left[\frac{r^4}{d+4} - \frac{\rho^4}{d} - \frac{2(d+2)}{d} \left(\frac{r^2}{d+2} - \frac{\rho^2}{d} \right) \rho^2 \right].$$

Thus $a_1(\rho) = 0$ for $\rho = \rho_1^* := \left(\frac{d}{d+2}\right)^{\frac{1}{2}} r$. And this is the only solution of the equation $a_1(\rho) = 0$. As a result we get the cubature formula

$$\int_{B(r)} u(x)\,dx = \frac{r}{d}\left(\frac{d+2}{d}\right)^{\frac{d-1}{2}} \int_{S(\rho_1^*)} u(x)\,d\sigma(x)$$

which is exact for all biharmonic functions u. This fact was established first by Dimitrov in [7].

Similarly, the equation $a_2(\rho) = 0$ has a unique solution

$$\rho_2^* = \left(\frac{d}{d+4}\right)^{\frac{1}{2}} r$$

and therefore, the cubature formula (based on integrals over the sphere $S^* := S(\rho_2^*)$)

$$\int_{B(r)} u(x)dx = \frac{r}{d}\left(\frac{d+4}{d}\right)^{\frac{d-1}{2}} \left\{\int_{S^*} u(x)d\sigma + \frac{r^2}{(d+2)(d+4)} \int_{S^*} \Delta u(x)d\sigma\right\}$$

integrates exactly all 3–harmonic functions u.

It would be interesting to prove the existence and uniqueness of the extremal radius ρ_{m-1}^* for all m.

Acknowledgement. The research was supported by the Bulgarian Ministry of Education under Grant No MM 802/98.

References

[1]   E. Almansi: *Sulle integrazione dell' equazione differenziale* $\Delta^{2m}u = 0$, Ann. Mat. Pura Appl., Suppl. 3, **2** (1898).

[2]   D. H. Armitage, S. J. Gardiner, W. Haußmann, L. Rogge: *Characterization of best harmonic and subharmonic L_1-approximations,* J. reine angew. Math. **478** (1996), 1–15.

[3]   N. Aronszajn, T. M. Creese, L. J. Lipkin: *Polyharmonic Functions,* Clarendon Press, Oxford 1983.

[4]   B. Bojanov: *An extension of the Pizzetti formula for polyharmonic functions,* Acta Math. Hungar. **91** (2001), 99–113.

[5]   B. Bojanov, D. Braess, N. Dyn: *Generalized Gaussian quadrature formulas,* J. Approx. Theory **48** (1986), 335–353.

[6]   B. D. Bojanov, D. K. Dimitrov: *Gaussian extended cubature formulae for polyharmonic functions,* Math. Comp. **70** (2001), 671–683.

[7]   D. K. Dimitrov: *Integration of polyharmonic functions,* Math. Comp. **65** (1996), 1269–1281.

[8]   K. F. Gauss: *Methodus nova integralium valores per approximationem,* Gotingensis Recentiores **2**, 1814; [Werke III, 1123–1162].

[9]   A. Ghizzetti, A. Ossicini: *Sull'esistenza e unicitat delle formule di quadrature Gaussiane,* Rend. Mat. **8** (1975), 1–15.

[10]   C. R. Hobby, J. R. Rice: *A moment problem in L_1-approximation,* Proc. Amer. Math. Soc. **16** (1965), 665–670.

[11]   S. Karlin: *Total Positivity,* Stanford University Press, California 1968.

[12]   M. Krein: *The ideas of P. L. Chebychev and A. A. Markov in the theory of limiting values of integrals and their further developments.* Uspekhi Fiz. Nauk (1951), 3–120 (in Russian); Amer. Math. Soc. Transl. Ser. 2, **12** (1951), 1–122.

[13]   I. P. Mysovskih: *Interpolatory Cubature Formulae,* Nauka, Moscow 1981 (in Russian).

[14]   M. Nicolescu: *Les Fonctions Polyharmoniques,* Hermann, Paris 1936.

[15]   P. Pizzetti: *Sulla media dei valore che una funzione dei punti della spazio assume alla superticie di una sfera,* Rendiconti Lincei, serie V, XVIII; 1 sem. (1909), 182–185.

Address:
Borislav Bojanov
Department of Mathematics
University of Sofia
5 James Bourchier Blvd.
BG–1164 Sofia
Bulgaria

International Series of Numerical Mathematics
Vol. 137, ©2001 Birkhäuser Verlag Basel/Switzerland

On the Structure of Kergin Interpolation for Points in General Position

Len Bos and Shayne Waldron

Abstract

For $n + 1$ points in $\mathbb{R}^d$, in general position, the Kergin polynomial interpolant of C^n functions may be extended to an interpolant of C^{d-1} functions. This results in an explicit set of reduced Kergin functionals naturally stratified by their dependence on certain directional derivatives of order k, $0 \leq k \leq d - 1$. We show that the polynomials dual to the functionals depending on derivatives of order k are multi–ridge functions of $d - k$ variables and moreover, that the polynomials dual to the purely interpolating functionals ($k = 0$) are always harmonic.

1 Introduction

Kergin interpolation is a natural multivariate extension of the Newton form of classical Hermite polynomial interpolation, introduced by Kergin in [4], and since then much studied (see, e.g., [6], [7], [1], [2]). It is most easily defined as follows.

Suppose that Θ is a sequence of $n + 1$ (possibly coincident) points in $\mathbb{R}^d$

$$\Theta := [\theta_0, \theta_1, \ldots, \theta_n].$$

Then the *Kergin interpolant* of $f \in C^n(\mathbb{R}^d)$ is the polynomial $K_\Theta f$ of degree n, defined by

$$K_\Theta f(x) := \sum_{k=0}^{n} \int_{[\theta_0,\ldots,\theta_k]} d^k f(x - \theta_0, x - \theta_1, \ldots, x - \theta_{k-1}). \qquad (1.1)$$

Here $d^k f$ denotes the k–th total derivative of f, and the linear functional

$$f \mapsto \int_{[\theta_0,\ldots,\theta_k]} f := \frac{1}{k! \operatorname{vol}_k(S)} \int_S f \circ A, \qquad (1.2)$$

where S is any k–simplex in $\mathbb{R}^s$ with (k–dimensional) volume $\text{vol}_k(S)$, and $A : \mathbb{R}^s \to \mathbb{R}^d$ is any affine map taking the $k+1$ vertices of S onto the points $\theta_0, \ldots, \theta_k$. The change of variables formula shows that (1.2) does not depend on the particular choice of S and A.

In the univariate case (1.1) reduces to the Newton form of the Hermite interpolant of f at Θ, via the Hermite–Genocchi formula for the divided difference of f at $\theta_0, \ldots, \theta_k$, i.e.,

$$[\theta_0, \ldots, \theta_k]f = \int_{[\theta_0, \ldots, \theta_k]} D^k f.$$

Often it is convenient to write $K_\Theta f$ in terms of directional derivatives

$$K_\Theta f(x) = \sum_{k=0}^{n} \int_{[\theta_0, \ldots, \theta_k]} D_{x-\theta_0} D_{x-\theta_1} \cdots D_{x-\theta_{k-1}} f,$$

and to make a particular choice of S and A in (1.2), such as

$$\int_{[\theta_0, \ldots, \theta_k]} f = \int_0^1 \int_0^{s_1} \cdots \int_0^{s_{k-1}} f(\theta_0 + s_1(\theta_1 - \theta_0) + \cdots + s_k(\theta_k - \theta_{k-1})) \, ds_k \cdots ds_2 \, ds_1.$$

Amongst others, Kergin interpolation enjoys the following desirable properties:

Interpolation at Θ: $K_\Theta f(\theta_j) = f(\theta_j)$, $0 \le j \le n$.

Symmetry in Θ: K_Θ does not depend on the ordering of the points in Θ.

Error Formula: For each $f \in C^{n+1}(\mathbb{R}^d)$,

$$f(x) - K_\Theta f(x) = \int_{[\theta_0, \ldots, \theta_n]} D_{x-\theta_0} D_{x-\theta_1} \cdots D_{x-\theta_n} f.$$

In particular, K_Θ is a linear projector onto $\Pi_n(\mathbb{R}^d)$, the polynomials on $\mathbb{R}^d$ of degree $\le n$.

Continuity and Coalescence: For each $f \in C^n(\mathbb{R}^d)$, the map

$$\mathbb{R}^{d \times (n+1)} \to \Pi_n(\mathbb{R}^d) : \Theta \mapsto K_\Theta f$$

is continuous, and if $\theta_0 = \cdots = \theta_n = a$, then $K_\Theta f$ is the degree n Taylor polynomial of f centred at a.

Rolle's Property: For *any* homogeneous constant coefficient partial differential operator $q(D)$ of degree $k \leq n$,

$$\int_{[\theta_0,\dots,\theta_k]} q(D)(f - K_\Theta f) = 0. \tag{1.3}$$

PDE Faithfulness: For *any* homogeneous constant coefficient partial differential operator $q(D)$ of degree $k \leq n$,

$$q(D)f = 0 \Rightarrow q(D)(K_\Theta f) = 0.$$

The last property involves functions $f : \mathbb{R}^d \to \mathbb{R}$ which are constant on the translates of some affine subspace V of $\mathbb{R}^d$ of codimension m. Since the collection of translates (also called cosets) $\mathbb{R}^d/V := \{x + V : x \in \mathbb{R}^d\}$ partitions $\mathbb{R}^d$,

f is constant on the translates of V

$\Longleftrightarrow$ for every affine map $A : \mathbb{R}^d \to \mathbb{R}^s$ with kernel a translate of V, $f = g \circ A$

$\Longleftrightarrow$ for some affine map $A : \mathbb{R}^d \to \mathbb{R}^s$ with kernel a translate of V, $f = g \circ A$

where in each case $g : \mathrm{ran}(A) \to \mathbb{R}$ is (well) defined on the range of A by $g(Ax) := f(x)$. We will refer to such an f as a *m–ridge or multi–ridge function*. If $m = 1$, and A is taken to be a nonzero linear functional $\lambda : x \mapsto \langle \lambda^*, x \rangle$, $\lambda^* \in \mathbb{R}^d$, then we obtain $f = g \circ \lambda$, which is commonly referred as a *ridge function*, or *plane wave*.

Ridge Friendliness: If $f = g \circ A$ is a multi–ridge function, where $A : \mathbb{R}^d \to \mathbb{R}^s$ is an affine map, and $g \in C^n(\mathrm{ran}(A))$, then

$$K_\Theta f = (K_{A\Theta} g) \circ A.$$

Because of the possibility of coalescence to a single point (Taylor interpolation), the Kergin interpolant is defined in general only for C^n functions. However, in the univariate case there is no smoothness requirement for *distinct* interpolation points, i.e., one has Lagrange interpolation to functions from $C(\mathbb{R})$.

This phenomenon extends to higher dimensional Kergin interpolants for points in general position in $\mathbb{R}^d$, which can be defined for functions of smoothness $f \in C^{d-1}(\mathbb{R}^d)$. This extension can be described by a reduced set of linear functionals (interpolation conditions) that depend only on certain directional derivatives up to order $d - 1$, and their corresponding dual polynomials from $\Pi_n(\mathbb{R}^d)$. A complete description of these reduced functionals is given in Theorem 2.3. It is natural to stratify them by the order of the derivative which they depend on, an "ascending" scale of complexity. In Theorem 2.4, we show the corresponding system of dual polynomials has a "descending" scale of complexity; precisely, that the dual polynomials to functionals involving derivatives of order $k = 0, \ldots, d - 1$, are $(d - k)$-ridge functions, i.e., functions of only $d - k$ variables. Finally, in Theorem 3.3 we show the surprising fact that the dual polynomials for the $n + 1$ point evaluation functionals are always harmonic.

The constant coefficient differential operators $q(D)$ on $\mathbb{R}^d$ used below are homogeneous of degree $r \geq 0$, and are described in terms of the (homogeneous) polynomials $q \in \Pi_r^0(\mathbb{R}^d)$ that correspond to them.

Lemma 1.1. *Suppose that $A : \mathbb{R}^d \to \mathbb{R}^s$ is an affine map, with linear part $L := A - A(0)$. If f is a multi–ridge function of the form $f = g \circ A$, $g \in C^r(\mathrm{ran}(A))$, then*

$$\int_\Theta q(D)f = \int_{A\Theta} (q \circ L^*)(D)g, \qquad \text{for all } q \in \Pi_r^0(\mathbb{R}^d), \tag{1.4}$$

where $L^* : \mathbb{R}^s \to \mathbb{R}^d$ is the adjoint of L.

Proof. We compute

$$\begin{aligned}
D_u f(x) &:= \lim_{t \to 0} \frac{g(A(x + tu)) - g(A(x))}{t} \\
&= \lim_{t \to 0} \frac{g(Ax + tLu) - g(A(x))}{t} \\
&= (D_{Lu}g)(Ax),
\end{aligned}$$

and so (by iterating)

$$D_{u_1} \cdots D_{u_r} f = (D_{Lu_1} \cdots D_{Lu_r} g) \circ A, \qquad \text{for all } u_1, \ldots, u_r \in \mathbb{R}^d.$$

Hence it follows from (1.2) that

$$\int_\Theta D_{u_1} \cdots D_{u_r} f = \int_\Theta (D_{Lu_1} \cdots D_{Lu_r} g) \circ A = \int_{A\Theta} D_{Lu_1} \cdots D_{Lu_r} g. \tag{1.5}$$

Since the differential operator $q(D) = D_{u_1} \cdots D_{u_r}$ corresponds to the homogeneous polynomial $q(x) := (u_1^T x) \cdots (u_r^T x)$, and $D_{Lu_1} \cdots D_{Lu_r}$ to

$$((Lu_1)^T x) \cdots ((Lu_r)^T x) = (u_1^T L^* x) \cdots (u_r^T L^* x) = q(L^* x) = (q \circ L^*)(x),$$

(1.5) can be expressed in the form (1.4). $\qquad \square$

2 The Ridge Structure of the Kergin Dual Polynomials

Definition 2.1. *Suppose that V is a linear subspace of $\mathbb{R}^d$, with orthogonal complement $V^\perp$. Let $\Pi_k(V)$ denote the subspace of $\Pi_k(\mathbb{R}^d)$ consisting of multi-ridge polynomials which vary in the V directions, i.e., which are constant on translates of $V^\perp$, and $\Pi_k^0(V)$ those which are in addition homogeneous of degree k.*

If V has dimension m, then the natural identification with polynomials on $\mathbb{R}^m$ gives

$$\dim(\Pi_k(V)) = \binom{k+m}{m}, \qquad \dim(\Pi_k^0(V)) = \binom{k+m-1}{m-1}.$$

It follows from the Rolle's property and the error formula that K_Θ is the unique linear projector of $C^n(\mathbb{R}^d)$ onto $\Pi_n(\mathbb{R}^d)$ for which $\lambda(K_\Theta f) = \lambda(f)$ for all $\lambda \in \Lambda$, where

$$\Lambda := \sum_{j=1}^{n+1} \sum_{\substack{\Psi \subset \Theta \\ \#\Psi = j}} \operatorname{Span}\{f \mapsto \int_\Psi q(D)f : q \in \Pi_{j-1}^0(\mathbb{R}^d)\}.$$

Here $\Psi \subset \Theta$, $\#\Psi = j$ indicates that Ψ is a subsequence of Θ of cardinality j. Henceforth we will refer to Λ as the (space of) interpolation conditions of K_Θ.

Definition 2.2. *For Ψ a sequence of $k+1$ affinely independent points in $\mathbb{R}^d$, let Λ_Ψ^i be the space of (continuous) linear functionals $C^i(\operatorname{conv}(\Psi)) \to \mathbb{R}$*

$$\Lambda_\Psi^i := \{f \mapsto \int_\Psi q(D)f : q \in \Pi_i^0(\operatorname{aff}(\Psi)^\perp)\}, \qquad i = 0, 1, \dots. \tag{2.1}$$

The $q(D)$ occuring above consist of all i–th order derivatives normal to $\operatorname{aff}(\Psi)$, i.e., are spanned by $D_{n_1} D_{n_2} \cdots D_{n_i}$, where the $n_1, n_2, \dots, n_i$ are directions orthogonal to $\operatorname{aff}(\Psi)$.

The following decomposition of the interpolation conditions Λ was given in [8].

Theorem 2.3. (Reduced functionals) *Suppose the points in Θ are in general position. Then the space of interpolation conditions for K_Θ can be written as a direct sum*

$$\Lambda = \bigoplus_{j=1}^{d} \Lambda_{j-1}, \qquad \Lambda_{j-1} := \bigoplus_{\substack{\Psi \subset \Theta \\ \#\Psi = j}} \Lambda_\Psi \tag{2.2}$$

where the linear functionals

$$\Lambda_\Psi := \Lambda_\Psi^{j-1} = \{f \mapsto \int_\Psi q(D)f : q \in \Pi_{j-1}^0(\mathrm{aff}(\Psi)^\perp)\}$$

are continuous on $C^{d-1}(\mathbb{R}^d)$.

This was proved by using the identity

$$\int_{[\Theta,v,w]} D_{v-w}f = \int_{[\Theta,v]} f - \int_{[\Theta,w]} f,$$

to reduce the order of the derivatives occuring in the interpolation conditions as much as possible, then doing a dimension count of the conditions so obtained. This result implies K_Θ has a continuous extension to $C^{d-1}(\mathbb{R}^d) \to \Pi_n(\mathbb{R}^d)$.

Theorem 2.4. (Dual polynomials) *Let* P_Ψ^Θ *be the subspaces of* $\Pi_n(\mathbb{R}^d)$ *dual to* (2.2), *i.e., satisfying* $\dim(\Lambda_\Psi|_{P_\Psi^\Theta}) = \dim(P_\Psi^\Theta)$, *and* $\Lambda_{\Psi_1}(P_{\Psi_2}^\Theta) = 0$, $\Psi_1 \neq \Psi_2$, *which implies*

$$\Pi_n(\mathbb{R}^d) = \bigoplus_{j=1}^d \bigoplus_{\substack{\Psi \subset \Theta \\ \#\Psi = j}} P_\Psi^\Theta. \tag{2.3}$$

Then P_Ψ^Θ, $\#\Psi = j$, *consists of* $(d-j+1)$*-ridge functions in the directions* $\mathrm{aff}(\Psi)^\perp$, *i.e.,*

$$P_\Psi^\Theta \subset \Pi_n(\mathrm{aff}(\Psi)^\perp).$$

Proof. Fix $\Psi_0 \subset \Theta$, $\#\Psi_0 = k+1$, and consider the space of functionals Λ_{Ψ_0}. Let $A : \mathbb{R}^d \to \mathbb{R}^d$ be the orthogonal projection onto $V := \mathrm{aff}(\Psi_0)^\perp$. Then, by the ridge–friendliness of Kergin interpolation,

$$K_\Theta(g \circ A) = (K_{A\Theta}g) \circ A, \qquad \text{for all } g \in C^{d-1}(V). \tag{2.4}$$

Here $K_{A\Theta}g$ is the Kergin interpolant for the sequence of "projected" points $A\Theta \subset V$, which is defined for $g \in C^{d-1}(V)$ via (1.1). The $k+1$ points of Ψ_0 project onto a single point, say ψ_0, and the remaining projected points are distinct (since Θ is in general position).

By Lemma 1.1, and $A = A^*$ (orthogonal projections are self adjoint), the linear functionals in Λ_Ψ restricted to multi–ridge functions $f = g \circ A$ have the form

$$f \mapsto \int_\Psi q(D)f = \int_{A\Psi}(q \circ A)(D)g, \qquad q \in \Pi_{j-1}^0(\mathrm{aff}(\Psi)^\perp). \tag{2.5}$$

Let $\Lambda^|_\Psi$ denote the corresponding space of projected interpolation conditions

$$C^{d-1}(V) \to \mathbb{R} : g \mapsto \int_{A\Psi}(q \circ A)(D)g, \qquad q \in \Pi^0_{j-1}(\mathrm{aff}(\Psi)^\perp).$$

Then, by (2.4), the interpolation conditions of $K_{A\Theta}$ are $\Lambda^| := \sum_\Psi \Lambda^|_\Psi$, and we now show

$$\Lambda^| = \Lambda^|_{\Psi_0} \bigoplus \sum_{\substack{\Psi \subset \Theta \\ \Psi \neq \Psi_0}} \Lambda^|_\Psi, \qquad \dim(\Lambda^|_{\Psi_0}|_{\Pi_n(\mathbb{R}^d)|_V}) = \dim(\Lambda_{\Psi_0}). \qquad (2.6)$$

Firstly, by (2.5), the space $\Lambda^|_{\Psi_0}$ consists of the linear functionals

$$g \mapsto \frac{1}{k!}q(D)g(\psi_0), \qquad q \in \Pi^0_k(V),$$

and so has

$$\dim(\Lambda^|_{\Psi_0}|_{\Pi_n(\mathbb{R}^d)|_V}) = \dim(\Pi^0_k(V)) = \dim(\Lambda_{\Psi_0}).$$

Let L be any orthogonal projection onto a 1–dimensional subspace of V which maps only the points Ψ_0 onto $L\psi_0 = L(A\psi_0)$. Since Θ is in general position all but a finite number of the possible L will have this property. By (1.4) and the Hermite–Genocchi formula, a linear functional from $\Lambda^|_\Psi$ applied to a ridge function of the form $f = g \circ L$ is a multiple of the divided difference of the univariate function $g \in C^{d-1}(\mathrm{ran}(L))$ at the points $L\Psi$. Hence, if

$$\lambda^| \in \Lambda^|_{\Psi_0} \bigcap \sum_{\substack{\Psi \subset \Theta \\ \Psi \neq \Psi_0}} \Lambda^|_\Psi,$$

then $\lambda^|(f) = 0$, since the only $L\Psi$, $\Psi \neq \Psi_0$ that involve the point $L\psi_0$ $k + 1$ times are those with $\Psi \supset \Psi_0$. By the fundamentality of such ridge functions $f = g \circ L$, we have $\lambda^| = 0$.

Let Q be the subspace of $\Pi_n(\mathbb{R}^d)|_V$ dual to the first summand of (2.6), i.e., satisfying

$$\dim(\Lambda^|_{\Psi_0}|_Q) = \dim(\Lambda_{\Psi_0}), \qquad \Lambda^|_\Psi(Q) = 0, \quad \text{for all } \Psi \neq \Psi_0.$$

Then $P^\Theta_{\Psi_0} = Q \circ A \subset \Pi_n(\mathrm{aff}(\Psi_0)^\perp)$, since

$$\dim(\Lambda_{\Psi_0}|_{Q \circ A}) = \dim(\Lambda^|_{\Psi_0}|_Q) = \dim(\Lambda_{\Psi_0}) \quad \text{for all } \Psi \neq \Psi_0$$

and

$$\Lambda_\Psi(Q \circ A) = \Lambda_\Psi^\downarrow(Q) = 0, \quad \text{for all } \Psi \neq \Psi_0.$$

□

Remark. Kergin interpolation can also be defined for points in $\mathbb{C}^d$, via an extension of the functional (1.2), cf. [1]. This inherits all the properties of real Kergin interpolation, in a natural way, and a simple check shows the dual polynomials for Kergin interpolation to points in general position in $\mathbb{C}^d$ have the multi–ridge structure of Theorem 2.4.

Example. Consider the dual functionals to Λ_{Ψ_0}, $\#\Psi_0 = d$. This space is 1–dimensional, with a basis given by

$$\lambda : f \mapsto \int_{\Psi_0} D_n^{d-1} f,$$

where n is a unit normal to the hyperplane $\text{aff}(\Psi_0)$. Let $p = p_\lambda \in P_{\Psi_0}^\Theta$ be the polynomial dual to λ, then by our proof, this is the ridge function

$$p(x) = g(<n, x>), \qquad g := (\cdot - \psi_0)^{d-1} \prod_{\theta \in \Theta \setminus \Psi_0} (\cdot - <n, \theta>), \quad \psi_0 := <n, \Psi_0> .$$

where g is the Hermite interpolant to the data

$$g(<n, \theta>) = 0, \quad \theta \in \Theta \setminus \Psi, \qquad\qquad (n - d \text{ conditions})$$

and

$$g(\psi_0) = Dg(\psi_0) = \cdots = D^{d-2}g(\psi_0) = 0, \quad \frac{D^{d-1}g(\psi_0)}{(d-1)!} = 1 \qquad (d \text{ conditions}).$$

3   Harmonicity of the Dual Polynomials for Point Evaluation

Let Δ denote the Laplace operator in $\mathbb{R}^d$. It is well known that the map

$$\Delta|_{\Pi_n(\mathbb{R}^d)} : \Pi_n(\mathbb{R}^d) \to \Pi_{n-2}(\mathbb{R}^d) : f \mapsto \Delta f$$

is onto, and so its kernel $\mathbb{H}_n(\mathbb{R}^d)$, the harmonic polynomials of degree $\leq n$ in $\mathbb{R}^d$, has

$$\dim(\mathbb{H}_n(\mathbb{R}^d)) = \binom{n+d}{d} - \binom{n-2+d}{d}.$$

Since $\dim(\mathbb{H}_n(\mathbb{R}^d)) < \dim(\Pi_n(\mathbb{R}^d)) = \dim \Lambda$, $n \geq 2$, the Kergin interpolation conditions Λ of (2.2) are not linearly independent over $\mathbb{H}_n(\mathbb{R}^d)$, i.e., as functionals restricted to $\mathbb{H}_n(\mathbb{R}^d)$. We now describe these dependencies, which also hold for all harmonic functions. These then imply that the dual polynomials to point evaluations are harmonic for $d > 1$.

For a sequence of points Ψ in $\mathbb{R}^d$, denote by $\Delta_{\Psi}^{\perp}$ the Laplacian in the directions orthogonal to $\mathrm{aff}(\Psi)$, i.e.,

$$\Delta_{\Psi}^{\perp} = q_{\Psi}^{\perp}(D) := D_{\mathrm{n}_1}^2 + D_{\mathrm{n}_2}^2 + \cdots + D_{\mathrm{n}_{d-k}}^2, \qquad q_{\Psi}^{\perp} \in \Pi_2^0(\mathrm{aff}(\Psi)^{\perp}),$$

where $\mathrm{n}_1, \ldots, \mathrm{n}_{d-k}$ is any orthonormal basis for $\mathrm{aff}(\Psi)^{\perp}$. Denote by $\mathrm{conv}(\Psi)$ the convex hull of the points in Ψ.

Lemma 3.1. *Suppose that Ψ is a sequence of $k + 1$ affinely independent points in $\mathbb{R}^d$, and let $\lambda \in \Lambda_{\Psi}^i$, $i \geq 2$, be the linear functional given by*

$$\lambda(f) := \int_{\Psi} \Delta_{\Psi}^{\perp} p(D) f, \qquad p \in \Pi_{i-2}^0(\mathrm{aff}(\Psi)^{\perp}). \tag{3.1}$$

Then

$$\lambda|_{\mathbb{H}} \in \sum_{\substack{\tilde{\Psi} \subset \Psi \\ \#\tilde{\Psi} = k}} \Lambda_{\tilde{\Psi}}^{i-1}|_{\mathbb{H}}, \qquad \mathbb{H} := \{f \in C^i(\mathrm{conv}(\Psi)) : \Delta f = 0\}. \tag{3.2}$$

Proof. Let Δ_{Ψ} denote the Laplacian in directions parallel to $\mathrm{aff}(\Psi)$, i.e., split $\Delta = \Delta_{\Psi} + \Delta_{\Psi}^{\perp}$. Suppose that f is harmonic. Then

$$\Delta p(D) f = p(D) \Delta f = 0 \Rightarrow \Delta_{\Psi}^{\perp} p(D) f = -\Delta_{\Psi} p(D) f,$$

and so, by (1.2),

$$\lambda(f) = \int_{\Psi} \Delta_{\Psi}^{\perp} p(D) f = -\int_{\Psi} \Delta_{\Psi} p(D) f = -\frac{1}{k! \, \mathrm{vol}_k(\mathrm{conv}(\Psi))} \int_{\mathrm{conv}(\Psi)} \Delta_{\Psi} p(D) f.$$

Now, by Green's theorem,

$$\int_{\mathrm{conv}(\Psi)} \Delta_{\Psi} g = \int_{\partial(\mathrm{conv}(\Psi))} D_{\mathrm{n}} g = \sum_{\substack{\tilde{\Psi} \subset \Psi \\ \#\tilde{\Psi} = k}} \int_{\mathrm{conv}(\tilde{\Psi})} D_{\mathrm{n}} g, \quad \text{for all } g \in C^2(\mathrm{conv}(\Psi)),$$

where n is the outward normal to the boundary of the k–simplex $\mathrm{conv}(\Psi)$, which consists of the faces $\mathrm{conv}(\tilde{\Psi})$. Hence, using (1.2), $\lambda|_{\mathbb{H}}$ can be expressed as a linear combination of functionals

$$\lambda(f) = \sum_{\substack{\tilde{\Psi} \subset \Psi \\ \#\tilde{\Psi} = k}} c_{\tilde{\Psi}} \int_{\tilde{\Psi}} D_{\mathrm{n}} p(D) f, \qquad c_{\tilde{\Psi}} := -\frac{\mathrm{vol}_{k-1}(\mathrm{conv}(\tilde{\Psi}))}{k \, \mathrm{vol}_k(\mathrm{conv}(\Psi))},$$

each of which belongs to $\Lambda_{\tilde{\Psi}}^{i-1}$, since $p \in \Pi_{i-2}^0(\mathrm{aff}(\Psi)^\perp) \Rightarrow p \in \Pi_{i-2}^0(\mathrm{aff}(\tilde{\Psi})^\perp)$.

$\square$

There are also linear dependencies, of a different type, between the functionals Λ_{Ψ}^1 over harmonic functions.

Lemma 3.2. *Suppose that Ψ is a sequence of $d+1$ affinely independent points in $\mathbb{R}^d$. Then, for $f \in C^1(\mathrm{conv}(\Psi))$ harmonic,*

$$\sum_{\substack{\tilde{\Psi} \subset \Psi \\ \#\tilde{\Psi}=d}} a_{\tilde{\Psi}} \int_{\tilde{\Psi}} D_{\mathrm{n}} f = 0, \qquad a_{\tilde{\Psi}} := (d-1)! \qquad \mathrm{vol}_{d-1}(\mathrm{conv}(\tilde{\Psi})),$$

where n *denotes the outward normal to* $\mathrm{conv}(\tilde{\Psi})$ *as a face of the simplex* $\mathrm{conv}(\Psi)$.

Proof. By Green's Theorem,

$$\sum_{\substack{\tilde{\Psi} \subset \Psi \\ \#\tilde{\Psi}=d}} a_{\tilde{\Psi}} \int_{\tilde{\Psi}} D_{\mathrm{n}} f = \sum_{\substack{\tilde{\Psi} \subset \Psi \\ \#\tilde{\Psi}=d}} \int_{\mathrm{conv}(\tilde{\Psi})} D_{\mathrm{n}} f = \int_{\partial(\mathrm{conv}(\Psi))} D_{\mathrm{n}} f = \int_{\mathrm{conv}(\Psi)} \Delta f = 0.$$

$\square$

In [2], using the identification of $\mathbb{R}^2$ with $\mathbb{C}$, it was shown that the dual polynomials to point evaluation for Kergin interpolation to points in general position in $\mathbb{R}^2$ are given by the real part of dual polynomials for univariate complex Lagrange interpolation to the points in $\mathbb{C}$, and so are harmonic. We now show this harmonicity extends to higher dimensions (including the odd dimensions!).

Theorem 3.3. (Harmonicity). *For Kergin interpolation to Θ in general position in $\mathbb{R}^d$, $d \geq 2$, the dual polynomials to point evaluation are harmonic, i.e., $\Delta P_{[\psi]}^{\Theta} = 0$, for all $\psi \in \Theta$.*

Proof. We begin by giving an alternative proof of the bivariate result, which is easily generalised to cover the so called scale of mean value interpolants (see, e.g., [8]).

In the bivariate case, the reduced interpolation conditions Λ of (2.2) consist of $n+1$ point evaluations Λ_0, and $\binom{n+1}{2}$ first derivative functionals Λ_1. By Lemma 3.2, for each triple of points $\Psi := [\psi_1, \psi_2, \psi_3] \subset \Theta$, the associated first derivative functionals have a linear dependency over harmonic functions given by

$$\|\psi_2 - \psi_1\| \int_{[\psi_1,\psi_2]} D_{\mathrm{n}} f + \|\psi_3 - \psi_2\| \int_{[\psi_2,\psi_3]} D_{\mathrm{n}} f + \|\psi_1 - \psi_3\| \int_{[\psi_3,\psi_1]} D_{\mathrm{n}} f = 0.$$

Since each pair of points is in a triple containing a given point θ_0, this implies the n first derivative functionals that involve θ_0 and another point span Λ_1 over $\mathbb{H}_n(\mathbb{R}^2)$. But,

$$\dim(\Lambda|_{\mathbb{H}_n(\mathbb{R}^2)}) = \dim(\mathbb{H}_n(\mathbb{R}^2)) = 2n + 1 = (n + 1) + n = \dim(\Lambda_0) + n,$$

and so these n functionals must be a basis for Λ_1 over $\mathbb{H}_n(\mathbb{R}^2)$, which when appended by the $(n + 1)$ point evaluations Λ_0 forms a dual basis for $\mathbb{H}_n(\mathbb{R}^2)$. Let $h \in \mathbb{H}_n(\mathbb{R}^2)$ be the harmonic polynomial which is 1 at a point ψ of Θ, 0 at the others, and is annihilated by the n first derivative functionals above, and hence all of Λ_1. Then this h satisfies the conditions that characterise the Kergin dual polynomial to point evaluation at ψ, which is therefore harmonic.

Now the case $d \geq 3$. The idea is the same, except we use the linear dependencies of Lemma 3.1. Let us count how many different dependencies this gives. Since Θ is in general position, for $2 \leq k < d$, each of the $\binom{n+1}{k+1}$ subsequences Ψ of $k + 1$ points is affinely independent, and the subspace $\Lambda_\Psi^{\mathrm{dep}}$ of $\Lambda_\Psi = \Lambda_\Psi^k$ of functionals of the form (3.1) has

$$\dim(\Lambda_\Psi^{\mathrm{dep}}) = \dim(\Pi_{k-2}^0(\mathrm{aff}(\Psi)^\perp)) = \binom{k - 2 + (d - k - 1)}{d - k - 1} = \binom{d - 3}{d - k - 1}.$$

This gives a total number of dependencies

$$\sum_{k=2}^{d-1} \binom{d - 3}{d - k - 1}\binom{n + 1}{k + 1} = \binom{n - 2 + d}{d}, \tag{3.3}$$

as the above sum represents the number of ways of choosing d objects from $n - 2 + d$ objects after having split them into two groups, one of size $d - 3$ and the other of size $n + 1$. Let Λ_{Ψ_0} be a complement of $\Lambda_\Psi^{\mathrm{dep}}$ in Λ_Ψ, i.e., write $\Lambda_\Psi = \Lambda_{\Psi_0} \oplus \Lambda_\Psi^{\mathrm{dep}}$. Then, by (3.2),

$$\Lambda_\Psi|_{\mathbb{H}} = \Lambda_{\Psi_0}|_{\mathbb{H}} + \Lambda_\Psi^{\mathrm{dep}}|_{\mathbb{H}} \subset \Lambda_{\Psi_0}|_{\mathbb{H}} + \Lambda_{k-1}|_{\mathbb{H}},$$

where $\mathbb{H} := \{f \in C^{d-1}(\mathrm{conv}(\Theta)) : \Delta f = 0\}$, and so the space of functionals

$$\Lambda^* := \Lambda_0 \oplus \Lambda_1 \oplus \bigoplus_{j=3}^{d} \bigoplus_{\substack{\Psi \subset \Theta \\ \#\Psi = j}} \Lambda_{\Psi_0},$$

spans Λ over $\mathbb{H}$. By (3.3),

$$\dim(\Lambda^*) = \binom{n + d}{d} - \binom{n - 2 + d}{d} = \dim(\mathbb{H}_n(\mathbb{R}^d)),$$

and so Λ^* is dual to $\mathbb{H}_n(\mathbb{R}^d)$. Let $h \in \mathbb{H}_n(\mathbb{R}^d)$ be the harmonic polynomial which is 1 at a point ψ of Θ, 0 at the others, and is annihilated by the other summands in the definition of Λ^*, and hence by $\Lambda_1, \ldots, \Lambda_{d-1}$. Then this h satisfies the conditions that characterise the Kergin dual polynomial to point evaluation at ψ, which is therefore harmonic. $\square$

Remark. The above result can be generalised to the mean value interpolation of [3]. Briefly, for points in general position (see [8]), there is a family of linear projectors

$$\mathcal{H}_\Theta^{(m)} : C^{n-m-1}(\mathbb{R}^d) \to \Pi_{n-m-1}(\mathbb{R}^d), \quad 0 \le m \le d - 1,$$

called the scale of mean value interpolations, determined by the interpolation conditions

$$\Lambda = \bigoplus_{j=m+1}^{d} \Lambda_{j-m-1}^{(m)} = \bigoplus_{j=m+1}^{d} \bigoplus_{\substack{\Psi \subset \Theta \\ \#\Psi = j}} \Lambda_\Psi^{j-m-1},$$

which contain Kergin interpolation as the special case $K_\Theta = \mathcal{H}_\Theta^{(0)}$. A modification of the above argument shows that for all mean value interpolation operators, except Hakopian interpolation, when $m = d - 1$, the dual polynomials to the functionals $\Lambda_0^{(m)}$ (which involve the function, but no derivatives) are harmonic.

References

[1]   M. Andersson, M. Passare: *Complex Kergin interpolation*, J. Approx. Theory, **64** (1991), 214–225.

[2]   L. Bos, J.–P. Calvi: *Kergin interpolants at the roots of unity approximate C^2 functions*, J. Anal. Math. **72** (1997), 203–221.

[3]   T. N. T. Goodman: *Interpolation in minimum semi–norm, and multivariate B–splines*, J. Approx. Theory **37** (1983), 212–223.

[4]   P. Kergin: *A natural interpolation of C^k functions*, J. Approx. Theory **29** (1980), 278–293.

[5]   V. Ya. Lin, A. Pinkus: *Fundamentality of ridge functions*, J. Approx. Theory **75(3)** (1993), 295–311.

[6]   C. A. Micchelli: *A constructive approach to Kergin interpolation in* $\mathbb{R}^k$: *multivariate B–splines and Lagrange interpolation,* Rocky Mountain J. Math. **10** (1980), 485–497.

[7]   C. A. Micchelli, P. Milman: *A formula for Kergin interpolation in* $\mathbb{R}^k$, J. Approx. Theory **29** (1980), 294–296.

[8]   S. Waldron: *Mean value interpolation for points in general position,* preprint

Addresses:

Len Bos
Department of Mathematics and Statistics
University of Calgary
Calgary, Alberta
Canada T2N 1N4

Shayne Waldron
Department of Mathematics
University of Auckland
Private Bag 92019
Auckland
New Zealand

International Series of Numerical Mathematics
Vol. 137, ©2001 Birkhäuser Verlag Basel/Switzerland

Frames Containing a Riesz Basis and Approximation of the Inverse Frame Operator

Ole Christensen and Alexander Lindner

Abstract

A frame in a Hilbert space $\mathcal{H}$ allows every element in $\mathcal{H}$ to be written as a linear combination of the frame elements, with coefficients called frame coefficients. Calculations of those coefficients and many other situations where frames occur, require knowledge of the inverse frame operator. But usually it is hard to invert the frame operator if the underlying Hilbert space is infinite dimensional. We introduce a method for approximation of the inverse frame operator using finite subsets of the frame. In particular this allows to approximate the frame coefficients (even in ℓ^2–sense) using finite–dimensional linear algebra. We show that the general method simplifies when the frame contains a Riesz basis.

1 Introduction

Let $\mathcal{H}$ be a separable Hilbert space with the inner product $< \cdot, \cdot >$ linear in the first entry.

Definition 1.1. $\{f_i\}_{i \in I} \subseteq \mathcal{H}$ *is a frame if there exist constants* $A, B > 0$ *such that*

$$A||f||^2 \leq \sum_{i \in I} | < f, f_i > |^2 \leq B||f||^2$$

for all $f \in \mathcal{H}$. A, B *are called frame bounds.*

Given a frame $\{f_i\}_{i \in I}$, the *frame operator* is defined by

$$S : \mathcal{H} \to \mathcal{H}, \quad Sf = \sum_{i \in I} < f, f_i > f_i.$$

S is bounded, invertible, and self–adjoint; this leads to the important *frame decomposition*:

$$f = SS^{-1}f = \sum_{i \in I} < f, S^{-1}f_i > f_i, \quad \text{for all } f \in \mathcal{H}. \tag{1.1}$$

For practical purposes, it is a problem that calculation of the frame coefficients $< f, S^{-1} f_i >$ requires inversion of S. In the following we develop methods for approximation of S^{-1} using finite subsets of $\{f_i\}_{i \in I}$.

For convenience, we describe the theory for a frame indexed by the natural numbers. Given a frame $\{f_i\}_{i=1}^{\infty}$, let $n \in \mathbb{N}$ and consider $\{f_i\}_{i=1}^{n}$, which is a frame for $\mathcal{H}_n := \text{span}\{f_i\}_{i=1}^{n}$. One can prove that the orthogonal projection P_n of $\mathcal{H}$ onto $\mathcal{H}_n$ is given by

$$P_n f = \sum_{i=1}^{n} < f, S_n^{-1} f_i > f_i, \quad f \in \mathcal{H},$$

where

$$S_n : \mathcal{H}_n \rightarrow \mathcal{H}_n, \quad S_n f = \sum_{i=1}^{n} < f, f_i > f_i.$$

Observe that S_n^{-1} (and hence P_n) can be found using finite–dimensional linear algebra.

Our starting point is the theorem below, which is proved in [3].

Theorem 1.2. *Let $\{f_i\}_{i=1}^{\infty}$ be a frame. Given $n \in \mathbb{N}$, let A_n denote a lower frame bound for $\{f_i\}_{i=1}^{n}$ (as frame for $\text{span}\{f_i\}_{i=1}^{n}$) and choose $m(n)$ such that*

$$\sum_{i=n+m(n)+1}^{\infty} | < f_j, f_i > |^2 \leq \frac{A_n}{n^2} \quad for \ j = 1, ..., n.$$

Let $V_n : \mathcal{H}_n \rightarrow \mathcal{H}_n$ denote the frame operator for the finite family $\{P_n f_i\}_{i=1}^{n+m(n)}$. Then

$$V_n^{-1} P_n f \rightarrow S^{-1} f, \quad for \ all \ f \in \mathcal{H}.$$

Theorem 1.2 demonstrates that S^{-1} can be approximated arbitrary well in the strong operator topology using finite–dimensional linear algebra. However, for practical calculation of V_n^{-1} it is desirable that $m(n)$ is not too large. But for most frames of practical interest (see section 3 and 4) one can prove that

$$A_n \rightarrow 0 \ as \ n \rightarrow \infty,$$

which forces $m(n)$ to be large. In the next section we show that Theorem 1.2 can be improved under an extra assumption.

2   Frames Containing a Riesz Basis

Recall that $\{f_i\}_{i\in I} \subseteq \mathcal{H}$ is a *Riesz basis* for $\mathcal{H}$ if $\overline{span}\{f_i\}_{i\in I} = \mathcal{H}$ and there exist constants $A, B > 0$ such that

$$A\sum |c_i|^2 \leq ||\sum c_i f_i||^2 \leq B\sum |c_i|^2$$

for all finite sequences $\{c_i\}$.

Given a Riesz basis $\{f_i\}_{i\in I}$, it is well known [8] that there exists a dual Riesz basis $\{g_i\}_{i\in I}$ such that

$$f = \sum_{i\in I} < f, g_i > f_i, \quad \text{for all } f \in \mathcal{H} \tag{2.1}$$

Observe the similarity between the equations (1.1) and (2.1)! In both cases the convergence is unconditional, i.e., independent of the order of summation. Intuitively, it is natural to think about a frame as an "overcomplete basis", but it turns out that this does not hold in the strict sense: There exist frames $\{f_i\}_{i\in I}$ for which no subfamily is a basis for $\mathcal{H}$. This is the reason for the following definition.

Definition 2.1. *A frame $\{f_i\}_{i\in I}$ contains a Riesz basis if there is an index set $J \subseteq I$ for which $\{f_i\}_{i\in J}$ is a Riesz basis for $\mathcal{H}$.*

General information about frames containing a Riesz basis can be found in [1] and [2].

For a frame containing a Riesz basis, Theorem 1.2 can be improved:

Theorem 2.2. *Let $\{f_i\}_{i=1}^{\infty}$ be a frame, containing a Riesz basis $\{f_i\}_{i\in J}$ with lower (Riesz) bound A. Choose finite index sets I_n for which*

$$I_1 \subseteq I_2 \subseteq ... \subseteq I_n \uparrow J,$$

and let P_n be the orthogonal projection onto $span\{f_i\}_{i\in I_n}$. Given $n \in \mathbb{N}$, choose a finite set J_n containing I_n such that

$$\sum_{i\notin J_n} | < f_j, f_i > |^2 \leq \frac{A}{n \cdot |I_n|}, \quad \text{for all } j \in I_n.$$

Let $V_n : \mathcal{H}_n \to \mathcal{H}_n$ denote the frame operator for the finite family $\{P_n f_i\}_{i\in J_n}$. Then

$$V_n^{-1} P_n f \to S^{-1} f \text{ as } n \to \infty, \quad \text{for all } f \in \mathcal{H}.$$

Proof. Let $n \in \mathbb{N}$. First, it can be proved that for all $f \in \mathcal{H}_n$,

$$\sum_{i \notin J_n} | < f, f_i > |^2 \leq \tfrac{1}{n} ||f||^2.$$

and

$$< (P_n S - V_n) f, f >= \sum_{i \notin J_n} | < f, f_i > |^2.$$

So $P_n S - V_n$ is a positive operator on $\mathcal{H}_n$ and $||(P_n S - V_n)_{|\mathcal{H}_n}|| \leq \tfrac{1}{n}$.

We leave it to the reader to prove that $A - \tfrac{1}{n}$ is a lower frame bound for $\{P_n f_i\}_{i \in J_n}$; this implies that $||V_n^{-1}|| \leq \frac{1}{A - \frac{1}{n}}$. Now, for $f \in \mathcal{H}$ we obtain that

$$||S^{-1} f - V_n^{-1} P_n f||$$

$$\leq \quad ||(I - P_n) S^{-1} f|| + ||P_n S^{-1} f - V_n^{-1} P_n f||$$

$$\leq \quad ||(I - P_n) S^{-1} f|| + ||V_n^{-1}|| \cdot ||V_n P_n S^{-1} f - P_n f||$$

$$\leq \quad ||(I - P_n) S^{-1} f|| + \frac{1}{A - \frac{1}{n}} (||V_n P_n S^{-1} f - P_n S P_n S^{-1} f||$$
$$+ \quad ||P_n S P_n S^{-1} f - P_n f||)$$

$$\leq \quad ||(I - P_n) S^{-1} f|| + \frac{1}{A - \frac{1}{n}} (||(V_n - P_n S) P_n S^{-1} f|| + ||S P_n S^{-1} f - f||)$$

$$\leq \quad ||(I - P_n) S^{-1} f|| + \frac{1}{A - \frac{1}{n}} (\frac{1}{n} \cdot ||P_n S^{-1} f|| + ||S|| \cdot ||P_n S^{-1} f - S^{-1} f||)$$

$$\leq \quad \frac{1}{n A (A - \frac{1}{n})} \cdot ||f|| + (\frac{B}{A - \frac{1}{n}} + 1) ||(I - P_n) S^{-1} f||. \qquad \square$$

The importance of Theorem 2.2 lies in the fact that the lower bound A_n appearing in Theorem 1.2 is replaced by A – independently of n. Furthermore, a typical value for A is ~ 1, while A_n is usually much smaller.

3 Frames of Exponentials

Let $\{\lambda_n\}_{n\in\mathbb{Z}} \subseteq \mathbb{R}$. A frame for $L^2(-\pi,\pi)$ of the form $\{e^{i\lambda_n x}\}_{n\in\mathbb{Z}}$ is called a *frame of exponentials*. This is actually the context in which frames were introduced in the original paper [6] by Duffin–Schaeffer in 1952.

Whether $\{e^{i\lambda_n x}\}_{n\in\mathbb{Z}}$ is a frame or not depends on the *density* of $\{\lambda_n\}_{n\in\mathbb{Z}}$. Given $r > 0$, let $n^-(r)$ denote the minimal number of points from $\{\lambda_n\}_{n\in\mathbb{Z}}$ to be found in an interval of length r and let

$$D^-(\{\lambda_n\}) := \lim_{r\to\infty} \frac{n^-(r)}{r}.$$

$D^-(\{\lambda_n\})$ is called the *lower Beurling density* of $\{\lambda_n\}_{n\in\mathbb{Z}}$.

Recall that $\{\lambda_n\}_{n\in\mathbb{Z}}$ is *separated* (with separation constant δ) if

$$|\lambda_n - \lambda_m| \geq \delta > 0 \text{ whenever } n \neq m.$$

$\{\lambda_n\}_{n\in\mathbb{Z}}$ is *relatively separated* if $\{\lambda_n\}_{n\in\mathbb{Z}}$ is a finite union of separated sets.

Seip [7] proved the following:

Theorem 3.1. *If $\{\lambda_n\}_{n\in\mathbb{Z}}$ is separated and has lower density strictly larger than one, then $\{e^{i\lambda_n x}\}_{n\in\mathbb{Z}}$ is a frame for $L^2(-\pi,\pi)$ containing a Riesz basis.*

It can be proved that if $\{e^{i\lambda_n x}\}_{n\in\mathbb{Z}}$ is a Riesz basis, then $D^-(\{\lambda_n\}) = 1$. For a regular distribution of points, i.e., $\lambda_n = nb$ for some $b > 0$, the lower density is $D^-(\{\lambda_n\}) = \frac{1}{b}$. When $b < 1$, it follows by Theorem 3.1 that $\{e^{inbx}\}_{n\in\mathbb{Z}}$ is a frame for $L^2(-\pi,\pi)$ containing a Riesz basis $\{e^{inbx}\}_{n\in J}$ for some $J \subseteq \mathbb{Z}$. It is interesting to observe that if b is irrational, no subfamily of the form $\{nNb\}_{n\in\mathbb{Z}}$ has density one; thus the Riesz basis $\{e^{inbx}\}_{n\in J}$ necessarily corresponds to points $\{nb\}_{n\in J}$ which are irregular distributed.

Seip also proved that for $\{\lambda_n\} = \{n(1 - |n|^{-1/2})\}_{|n|>1}$, $\{e^{i\lambda_n x}\}$ is a frame for $L^2(-\pi,\pi)$ which does not contain a Riesz basis.

We now return to the question about approximation of the inverse frame operator. First, it turns out to be very difficult to get good estimates for the lower frame bounds for a finite set of exponentials $\{e^{i\lambda_n x}\}_{n=1}^N$ in $L^2(-\pi,\pi)$. Assuming that $\{\lambda_n\}_{n=1}^N$ is separated with separation constant δ, it has been proved in [5] that

$$A_N := 1.6 \cdot 10^{-14} \cdot (\delta/2)^{2N+1} \cdot ((N+1)!)^{-8}$$

is a lower bound, but this bound is clearly too small for practical purposes. We would like to pose it as an open problem:

How can one obtain good estimates for the lower frame bound for a finite set of exponentials?

Without good estimates for the lower frame bound, Theorem 1.2 is not very useful for general frames of exponentials. The situation improves if $\{e^{i\lambda_n x}\}_{n\in\mathbb{Z}}$ contains a Riesz basis $\{e^{i\lambda_k x}\}_{k\in J}$. For convenience, assume that $\{\lambda_k\}_{k\in\mathbb{Z}}$ is separated and ordered such that

$$\cdots \lambda_{-1} \leq \lambda_0 \leq \lambda_1 \leq \cdots$$

Choose $I_1 \subseteq I_2 \subseteq \cdots \subseteq I_n \uparrow J$. For $n \in \mathbb{N}$, let $\tilde{n} := \max_{k\in I_n}|k|$ and let

$$J_n = \{k\}_{|k|\leq m(n)+\tilde{n}};$$

As before, P_n denotes the orthogonal projection onto $\mathrm{span}\{e^{i\lambda_k x}\}_{k\in I_n}$.

Theorem 3.2. *Suppose that the frame $\{e^{i\lambda_k x}\}_{k\in\mathbb{Z}}$ contains a Riesz basis $\{e^{i\lambda_k x}\}_{k\in J}$ with lower bound A. Given $n \in \mathbb{N}$, choose*

$$m(n) \geq \frac{8 \cdot n \cdot |I_n|}{\delta^2 A}$$

and let $V_n : \mathcal{H}_n \to \mathcal{H}_n$ denote the frame operator for the finite family $\{P_n e^{i\lambda_k x}\}_{k\in J_n}$. Then for all $f \in L^2(-\pi, \pi)$,

$$V_n^{-1} P_n f \to S^{-1} f, \quad as \ n \to \infty.$$

The proof can be found in [5].

4   Gabor Frames

Definition 4.1. *Let $g \in L^2(\mathbb{R})$, $a, b > 0$. A frame for $L^2(\mathbb{R})$ of the form*

$$\{e^{2\pi imbx}g(x - na)\}_{m,n\in\mathbb{Z}}$$

is called a Gabor frame.

By introducing the operators on $L^2(\mathbb{R})$

$$\text{Translation by } a \in \mathbb{R} : (T_a f)(x) = f(x - a), \ x \in \mathbb{R}$$

and

$$\text{Modulation by } b \in \mathbb{R} : (E_b g)(x) = e^{2\pi i b x} f(x), \ x \in \mathbb{R},$$

we can use $\{E_{mb} T_{na} g\}_{m,n \in \mathbb{Z}}$ as short notation for a Gabor frame.

Usually, one thinks about a Gabor frame $\{E_{mb} T_{na} g\}_{m,n \in \mathbb{Z}}$ as the set of time–frequency shifts of g along the *lattice* $\{(na, mb)\}_{m,n \in \mathbb{Z}} \subseteq \mathbb{R}^2$. In order for $\{E_{mb} T_{na} g\}_{m,n \in \mathbb{Z}}$ to be a frame, the lattice has to be dense enough, in the sense that $ab \leq 1$. Also, if $\{E_{mb} T_{na} g\}_{m,n \in \mathbb{Z}}$ is a frame, then

$$\{E_{mb} T_{na} g\}_{m,n \in \mathbb{Z}} \text{ is a Riesz basis} \Leftrightarrow ab = 1 \qquad (4.1)$$

It is easy to construct Gabor frames containing a Riesz basis. For example, by choosing g, a, b such that $\{E_{mb} T_{na} g\}_{m,n \in \mathbb{Z}}$ is a Riesz basis and letting $N \in \mathbb{N}$, the family $\{e^{i2\pi m b x} g(x - n \frac{a}{N})\}_{m,n \in \mathbb{Z}}$ is a frame containing a Riesz basis.

For another example, let $\{\lambda_m\}_{m \in I} \subseteq \mathbb{R}$ be a separated set for which $D^-(\{\lambda_m\}) > 1$. By Theorem 3.1, $\{e^{i\lambda_m x}\}_{m \in I}$ is a frame for $L^2(-\pi, \pi)$ containing a Riesz basis. Thus, for every function g for which

$$\text{supp}(g) \subseteq [-\pi, \pi], \quad A \leq |g(x)| \leq B, \quad a.e. \ x \in [-\pi, \pi],$$

the family $\{e^{i\lambda_m x} g(x - n2\pi)\}_{m \in I, n \in \mathbb{Z}}$ is a Gabor frame for $L^2(\mathbb{R})$ which contains a Riesz basis. The remark after Theorem 3.1 implies that even in the lattice case $\lambda_m = 2\pi m b$, only for very special values of b, the Riesz basis contained in $\{e^{i2\pi m b x} g(x - n2\pi)\}_{m,n \in \mathbb{Z}}$ will correspond to a sublattice of the form $\{e^{i2\pi m b' x} g(x - n2\pi)\}_{m,n \in \mathbb{Z}}$, where $b' > 0$.

However, besides such constructions, it is very difficult to decide when a given Gabor frame contains a Riesz basis. We put it as an open problem:

When does a Gabor frame $\{E_{mb} T_{na} g\}_{m,n \in \mathbb{Z}}$ contain a Riesz basis?

We now want to apply the general approximation theory to Gabor frames $\{E_{kb} T_{la} g\}_{k,l \in \mathbb{Z}}$ containing a Riesz basis $\{E_{kb} T_{la} g\}_{(k,l) \in J}$. First, choose finite sets $\{I_n\}$ such that

$$I_1 \subseteq I_2 \subseteq ... \subseteq I_n \uparrow J.$$

Given $n \in \mathbb{N}$, let $\tilde{n} = \max_{(k,l) \in I_n} \{|k|, |l|\}$ and let

$$J_n = \{(k,l) : |k|, |l| \leq \tilde{n} + m(n)\}.$$

As before, the question is how to choose $m(n) \in \mathbb{N}$ such that Theorem 2.2 can be applied. We need the Lemma below.

Lemma 4.2. *If $|k'|, |l'| \leq \tilde{n}$, then for all choices of $m(n) \in \mathbb{N}$,*

$$\sum_{(k,l) \notin J_n} | < E_{kb} T_{la} g, E_{k'b} T_{l'a} g > |^2 \leq \sum_{\{(k,l):|k|,|l| \leq m(n)\}^c} | < E_{kb} T_{la} g, g > |^2.$$

In particular, the estimate holds when $(k', l') \in I_n$.

Proof. Suppose that $|k'|, |l'| \leq \tilde{n}$. Then

$$\sum_{(k,l) \notin J_n} | < E_{kb} T_{la} g, E_{k'b} T_{l'a} g > |^2$$
$$= \sum_{(k,l) \notin J_n} | < E_{(k-k')b} T_{(l-l')a} g, g > |^2$$
$$= \sum_{k,l \in \mathbb{Z}} | < E_{(k-k')b} T_{(l-l')a} g, g > |^2$$
$$- \sum_{\{(k,l):|k|,|l| \leq \tilde{n}+m(n)\}} | < E_{(k-k')b} T_{(l-l')a} g, g > |^2.$$

Now, since $|k'|, |l'| \leq \tilde{n}$,

$$\sum_{\{(k,l):|k|,|l| \leq \tilde{n}+m(n)\}} | < E_{(k-k')b} T_{(l-l')a} g, g > |^2$$
$$\geq \sum_{\{(k,l):|k|,|l| \leq m(n)\}} | < E_{kb} T_{la} g, g > |^2,$$

from which the Lemma follows. $\square$

Now, with the above definition of J_n and with P_n being the orthogonal projection onto $\mathrm{span}\{E_{kb} T_{la} g\}_{(k,l) \in I_n}$ we get the following consequence of Theorem 2.2:

Theorem   4.3.   *Suppose that $\{E_{kb} T_{la} g\}_{k,l \in \mathbb{Z}}$ contains a Riesz basis $\{E_{kb} T_{la} g\}_{(k,l) \in J}$ with lower bound A. For $n \in \mathbb{N}$, choose a number $m(n)$ such that*

$$\sum_{\{(k,l):|k|,|l| \leq m(n)\}^c} | < E_{kb} T_{la} g, g > |^2 \leq \frac{A}{n|I_n|}.$$

Let $V_n : \mathcal{H}_n \rightarrow \mathcal{H}_n$ be the frame operator for $\{P_n E_{kb} T_{la} g\}_{(k,l) \in J_n}$. Then, as $n \rightarrow \infty$,

$$V_n^{-1} P_n f \rightarrow S^{-1} f, \quad \text{for all } f \in L^2(\mathbb{R}).$$

If a Gabor frame does not contain a Riesz basis, one has to use Theorem 1.2 instead of Theorem 4.3. However, then we need some lower frame bounds for finite Gabor systems. The following theorem gives some lower bounds for certain functions g which have one–sided bounded support. Since it does not present any extra difficulty to handle *irregular* systems $\{e^{2\pi i\lambda_m x}g(x-a_n)\}_{m,n=1}^{M,N}$, i.e. where λ_m and a_n are not necessarily of the form $\lambda_m = bm$ and $a_n = an$ for some $a, b > 0$, we will immediately do it for these systems.

Theorem 4.4. *Let $a_1, \ldots, a_N$ and $\lambda_1, \ldots, \lambda_M$ be two finite separated sequences of real numbers, the latter separated by $\varepsilon > 0$. Let $g \in L^2(\mathbb{R})$ be such that $\operatorname{supp} g \subset (-\infty, c]$ for some $c \in \mathbb{R}$, and suppose there is a non–degenerate interval $I \subset [c - \varepsilon, c]$ and a positive number d such that*

$$|g(x)| \geq d \quad \text{for all } x \in I.$$

Denote a lower bound for $\{e^{2\pi i\lambda_m x}\}_{m=1}^M$ in $L^2(I)$ by A, and an upper bound for $\{e^{2\pi i\lambda_m x}g(x)\}_{m=1}^M$ in $L^2(\mathbb{R})$ by B'. Then $\{e^{2\pi i\lambda_m x}g(x-a_n)\}_{m=1,n=1}^{M,N}$ is linearly independent with lower frame bound

$$A_N = d^2 A \left(\frac{d^2 A}{16 B'} \right)^{N-1}.$$

Proof. Without loss of generality we suppose $a_1 < \ldots < a_N$. Since a finite sequence is a Riesz basis for its linear span if and only if it is linearly independent, and since in that case the frame bounds and the Riesz bounds coincide, it suffices to show that

$$\left\| \sum_{n=1}^k \sum_{m=1}^M c_{mn} e^{2\pi i\lambda_m(\cdot)}g(\cdot - a_n) \right\|_{L^2(\mathbb{R})} \geq \sqrt{A_k \sum_{n=1}^k \sum_{m=1}^M |c_{mn}|^2} \qquad (4.2)$$

holds for all $k \in \{1, \ldots, N\}$ and all sequences $\{c_{mn}\}_{m=1,n=1}^{M,N}$ of complex scalars. We do this by induction on k:
For $k = 1$, we have

$$\left\| \sum_{m=1}^M c_{m1} e^{2\pi i\lambda_m(\cdot)}g(\cdot - a_1) \right\|_{L^2(\mathbb{R})} \geq \left\| \sum_{m=1}^M c_{m1} e^{2\pi i\lambda_m(\cdot)}g(\cdot - a_1) \right\|_{L^2(I+a_1)} =$$

$$\left\| \sum_{m=1}^M c_{m1} e^{2\pi i\lambda_m(\cdot + a_1)}g \right\|_{L^2(I)} \geq d\sqrt{A \sum_{m=1}^M |c_{m1}|^2}.$$

Now suppose that $k \geq 2$ and that (4.2) holds for $k-1$. We distinguish between two cases:

Case 1:

$$\frac{1}{2}\sqrt{A_{k-1}\sum_{n=1}^{k-1}\sum_{m=1}^{M}|c_{mn}|^2} \geq \sqrt{B'\sum_{m=1}^{M}|c_{mk}|^2} \qquad (4.3)$$

We then have

$$\left\|\sum_{n=1}^{k}\sum_{m=1}^{M}c_{mn}e^{2\pi i\lambda_m(\cdot)}g(\cdot-a_n)\right\|_{L^2(\mathbb{R})} \geq$$

$$\left\|\sum_{n=1}^{k-1}\sum_{m=1}^{M}c_{mn}e^{2\pi i\lambda_m(\cdot)}g(\cdot-a_n)\right\|_{L^2(\mathbb{R})} - \left\|\sum_{m=1}^{M}c_{mk}e^{2\pi i\lambda_m(\cdot)}g(\cdot-a_k)\right\|_{L^2(\mathbb{R})} \geq$$

$$\sqrt{A_{k-1}\sum_{n=1}^{k-1}\sum_{m=1}^{M}|c_{mn}|^2} - \sqrt{B'\sum_{m=1}^{M}|c_{mk}|^2} \overset{(4.3)}{\geq}$$

$$\frac{1}{2}\sqrt{A_{k-1}\sum_{n=1}^{k-1}\sum_{m=1}^{M}|c_{mn}|^2} \overset{(4.3)}{\geq}$$

$$\frac{1}{4}\sqrt{A_{k-1}\sum_{n=1}^{k-1}\sum_{m=1}^{M}|c_{mn}|^2} + \frac{1}{2}\sqrt{B'\sum_{m=1}^{M}|c_{mk}|^2} \geq$$

$$\frac{1}{4}\sqrt{A_{k-1}}\sqrt{\sum_{n=1}^{k}\sum_{m=1}^{M}|c_{mn}|^2} \geq \sqrt{A_k\sum_{n=1}^{k}\sum_{m=1}^{M}|c_{mn}|^2}.$$

Case 2:

$$\frac{1}{2}\sqrt{A_{k-1}\sum_{n=1}^{k-1}\sum_{m=1}^{M}|c_{mn}|^2} \leq \sqrt{B'\sum_{m=1}^{M}|c_{mk}|^2} \qquad (4.4)$$

Then we have

$$\left\|\sum_{n=1}^{k}\sum_{m=1}^{M}c_{mn}e^{2\pi i\lambda_m(\cdot)}g(\cdot-a_n)\right\|_{L^2(\mathbb{R})} \geq$$

$$\left\|\sum_{n=1}^{k}\sum_{m=1}^{M}c_{mn}e^{2\pi i\lambda_m(\cdot)}g(\cdot-a_n)\right\|_{L^2(I+a_k)} \geq$$

$$\left\|\sum_{m=1}^{M} c_{mk} e^{2\pi i \lambda_m(\cdot)} g(\cdot - a_k)\right\|_{L^2(I+a_k)} \cdot \quad -$$

$$\sum_{n=1}^{k-1}\left\|\sum_{m=1}^{M} c_{mn} e^{2\pi i \lambda_m(\cdot)} g(\cdot - a_n)\right\|_{L^2(I+a_k)} \geq$$

$$d \sqrt{A \sum_{m=1}^{M} |c_{mk}|^2} \overset{(4.4)}{\geq} d\sqrt{A}\left(\frac{1}{2}\sqrt{\sum_{m=1}^{M} |c_{mk}|^2} + \frac{1}{4}\sqrt{\frac{A_{k-1}}{B'}\sum_{n=1}^{k-1}\sum_{m=1}^{M} |c_{mn}|^2}\right) \geq$$

$$\frac{d\sqrt{AA_{k-1}}}{4\sqrt{B'}}\sqrt{\sum_{n=1}^{k}\sum_{m=1}^{M} |c_{mn}|^2} = \sqrt{A_k \sum_{n=1}^{k}\sum_{m=1}^{M} |c_{mn}|^2},$$

thus completing the induction step. The proof is over. $\qquad\square$

With the help of Theorems 1.2, 4.4 and Lemma 4.2, one can now state an approximation result similar to Theorem 4.3, for certain Gabor frames which do not necessarily contain a Riesz basis. We leave the details to the reader.

Remark 4.5.

a) Note that an explicit value for the occuring lower bound A of $\{e^{2\pi i \lambda_m x}\}_{m=1}^{M}$ in $L^2(I)$ has been given in Section 3. Also, without further assumptions we can use $B' = N \cdot \|g\|^2$. In case $\{e^{2\pi i \lambda_m x} g(x - a_n)\}_{m=1,n=1}^{M,N}$ is a subset of a frame $\{e^{2\pi i \lambda_m x} g(x - a_n)\}_{m=1,n=1}^{\infty}$ with upper bound B, we can use $B = B'$ independently of N.

b) Since the Fourier Transform of the function $e^{2\pi i \lambda_m x} g(x - a_n)$ is given by $e^{2\pi i \lambda_m a_n} e^{-2\pi i a_n y} \hat{g}(y - \lambda_m)$ and since $|e^{2\pi i \lambda_m a_n}| = 1$, $\{e^{2\pi i \lambda_m x} g(x - a_n)\}_{m=1,n=1}^{M,N}$ is linearly independent if and only if $\{e^{-2\pi i a_n y} \hat{g}(y - \lambda_m)\}_{m=1,n=1}^{M,N}$ is, and the lower frame bounds are the same. Thus it is clear that an analogous statement to Theorem 4.4 holds if suitable conditions are posed on $\hat{g}$ instead on g.

c) Theorem 4.4 does only cover functions g with one–sided bounded support. In [4] there have been obtained lower bounds for a more general class of funtions g, which e.g. also includes the Gaussian.

References

[1] P. Casazza, O. Christensen: *Frames containing a Riesz basis and preservation of this property under perturbation.* SIAM J. Math. Anal. **29** (1998), 266–278.

[2] O. Christensen: *Frames containing a Riesz basis and approximation of the frame coefficients using finite dimensional methods,* J. Math. Anal. Appl. **199** (1996), 256–270.

[3] O. Christensen: *Finite-dimensional approximation of the inverse frame operator and applications to Weyl–Heisenberg frames and wavelet frames,* J. Fourier Anal. Appl. **6** (2000), 79–91.

[4] O. Christensen, A. Lindner: *Lower frame bounds for finite wavelet and Gabor systems,* Approx. Theory and its Appl., to appear.

[5] O. Christensen, A. Lindner: *Frames of exponentials: lower frame bounds for finite subfamilies, and approximation of the inverse frame operator,* Linear Alg. Appl. **323** (2001), 117–130.

[6] R. J. Duffin, A. C. Schaeffer: *A class of nonharmonic Fourier series,* Trans. Amer. Math. Soc. **72** (1952), 341–366.

[7] K. Seip: *On the connection between exponential bases and certain related sequences in $L^2(-\pi, \pi)$,* J. Funct. Anal. **130** (1995), 131–160.

[8] R. Young: *An Introduction to Nonharmonic Fourier Series,* Academic Press, New York, 1980.

Addresses:

Ole Christensen
Department of Mathematics
Technical University of Denmark
DK–2800 Lyngby
Denmark

Alexander M. Lindner
Fachbereich Mathematik
Technische Universität München
D–80290 München
Germany

International Series of Numerical Mathematics
Vol. 137, ©2001 Birkhäuser Verlag Basel/Switzerland

Kronecker Type Convolution of Function Vectors with one Refinable Factor

Costanza Conti

Abstract

The aim of this paper is the further investigation of some of the properties being preserved when convolving two function vectors. This work follows the investigations in [6] of convolved refinable function vectors *i.e.*, function vectors which are solutions of matrix refinement equations. Here, going further, we consider also the case where only one factor of the convolution is refinable showing how the properties of the unique refinable vector can be still used to define a subdivision–like algorithm.

1   Introduction

When approximating a curve or a surface, it is quite common to employ convolution to improve the smoothness and the order of approximation of a function or a family of functions used to represent the curve or the surface. A well known example is the B–splines of degree m with the uniform knots on $[0, m+1]$ defined by repeated convolution of characteristic functions $\chi_{[0,1)}$ (see, for instance, [3]). Other examples are given in [11], [15] where repeated convolutions are used to define new families of spline functions on the three and four directional mesh of the plane with increasing smoothness, degree and approximation order. In [5] convolution is used with an analogous goal in case of refinable function vectors on the four directional mesh of the plane.

An investigation of the properties preserved by convolution when dealing with shift invariant spaces generated by convolved refinable function vectors is given in [6] and [7]. In particular, there has been presented a detailed and systematic analysis of the convolution process proving that convolution of refinable function vectors provides a refinable function vector. Furthermore, it has been demonstrated that whenever each of the convolving factors has an associated convergent subdivision scheme the convolution produces a scheme which is convergent as well. Moreover, in [6], the approximation order of a convolved shift invariant space in terms of those of the two convolution factors is also discussed.

However, it is not always the case that both function vectors are refinable. Therefore, in this paper, we also take convolutions of a non–refinable function vector with a refinable function vector into account. An analogous investigation has been done with an application to image data compression in [2] where it has been shown how to get a remarkable compression rate on a shift invariant space generated by convolution relying on the unique factor which is refinable.

Here, instead, we focus our attention on the efficient algorithms of representing curves and surfaces in the shift–invariant spaces generated by convolution and discuss a "subdivision–like" method using the subdivision scheme we can define with respect to the refinable factor of the convolution. Furthermore, in case of complete refinability, we also show how to split the convolved subdivision scheme in two parts and thus we reduce the computational cost.

The paper is organized as follows: in Section 2, the problem is posed along with the discussion of a subdivision splitting. In Section 3, a method is presented for the non–complete refinable case and some convergence results are discussed. Finally, in Section 4, an example of a convolution combining a non–refinable and a refinable vector taken from [15] is given.

2   Refinability and Subdivision Splitting

Let Φ and Ψ be two given function vectors, of dimension n and m respectively, of d–variate compactly supported functions. We use the following 'Kronecker type' notation for the convolution of two vectors to define a new function vector of d–variate compactly supported functions:

$$\Theta := \Phi \divideontimes \Psi := \begin{pmatrix} \phi_1 \divideontimes \Psi \\ \phi_2 \divideontimes \Psi \\ \vdots \\ \phi_n \divideontimes \Psi \end{pmatrix}. \tag{2.1}$$

The convolution of a scalar function ϕ_i with the function vector Ψ is taken componentwise. This operation produces a function vector $\Theta = (\theta_1, \theta_2, \ldots, \theta_{mn})^T$ with $m \cdot n$ components of type $\theta_{(i-1)m+j} := \phi_i * \psi_j$, $i = 1, \ldots, n$, $j = 1, \ldots, m$.

It is worthwhile to note that this operation is not commutative. On the other hand, there exists a permutation matrix P such that $P(\Phi \divideontimes \Psi) = \Psi \divideontimes \Phi$.

In this paper we assume that at least one of the two factors of the convolution is refinable. We recall that a vector $\Psi = (\psi_1, \psi_2, \ldots, \psi_m)^T$ is *refinable*, if it satisfies a refinement equation

$$\Psi = \sum_{\alpha \in \mathbb{Z}^d} B_\alpha \, \Psi(2 \cdot -\alpha) \,, \tag{2.2}$$

where the *refinement (matrix) mask* $\mathbf{B} = (B_\alpha)_{\alpha \in \mathbb{Z}^d}$ is a matrix sequence of real $m \times m$ matrices with compact support. The *subdivision operator* is the linear operator $S_B : (\ell(\mathbb{Z}^d))^m \to (\ell(\mathbb{Z}^d))^m$ such that

$$(S_B \, \Lambda)_\alpha := \sum_{\beta \in \mathbb{Z}^d} B^T_{\alpha - 2\beta} \, \Lambda_\beta, \quad \alpha \in \mathbb{Z}^d \,, \tag{2.3}$$

where $(\ell(\mathbb{Z}^d))^m$ denotes the linear space of m dimensional vector sequences indexed by $\mathbb{Z}^d$. The *subdivision scheme* consists of the iterates of S_B for a given initial vector sequence $\Lambda \in (\ell(\mathbb{Z}^d))^m$:

> For a given initial vector sequence $\Lambda \in (\ell(\mathbb{Z}^d))^m$
> Put $\Lambda^{(0)} := \Lambda$ and
> Compute $\Lambda^{(k+1)} := S_B \, \Lambda^{(k)}, \; k = 0, 1, \ldots$.

Denoting with $(\ell^\infty(\mathbb{Z}^d))^m$ the linear subspace of all bounded m dimensional vector sequences indexed by $\mathbb{Z}^d$, the subdivision scheme *converges* for $\Lambda = (\lambda_1, \ldots, \lambda_m)^T \in (\ell^\infty(\mathbb{Z}^d))^m$, if there exists a continuous vector-valued function $f_\Lambda : \mathbb{R}^d \to \mathbb{R}^m$ ($f_\Lambda \neq 0$ for at least some initial data Λ) such that

$$\lim_{k \to \infty} \left\| \mathbf{f}_\Lambda^{(k)} - \Lambda^{(k)} \right\|_\infty = 0 \,. \tag{2.4}$$

The symbol $\mathbf{f}_\Lambda^{(k)}$ denotes the vector valued sequence $\left(f_\Lambda\left(\frac{\alpha}{2^k}\right) \right)_{\alpha \in \mathbb{Z}^d}$ while the symbol $\left\| \cdot \right\|_\infty$ denotes the infinity norm of a vector sequence defined as $\left\| \Lambda \right\|_\infty := \max_{i=1,\ldots,m} \left\| \lambda_i \right\|_\infty$.

Concerning matrix refinement equations and vector subdivision schemes we refer the reader to relevant literature like [4], [9], [12], [13]. The following theorem can be proved along the same lines as for the monodimensional case given in [4].

Theorem 2.1.   *Let the subdivision scheme associated with the subdivision operator S_B defined in (2.3) be convergent. Then there exists a row vector v such that*

$$v \sum_{\beta \in \mathbb{Z}^d} B_{\alpha - 2\beta} = v \; , \; \text{for all } \alpha \in E \; ,$$

where E is the set of vertices of $[0,1]^d$. Moreover, let V_Ψ be the space of all common left eigenvectors of $\sum_{\beta \in \mathbb{Z}^d} B_{\alpha - 2\beta}$ for all $\alpha \in E$, corresponding to the eigenvalue 1.

Then V_Ψ is also the space of left eigenvectors corresponding to the eigenvalue 1 of $H_\Psi(0)$, i.e., the matrix mask symbol

$$H_\Psi(\xi) := \frac{1}{2^d} \sum_{\alpha \in \mathbb{Z}^d} B_\alpha \, e^{-2\pi i \alpha \cdot \xi}$$

evaluated at 0.

In addition, if $l = \dim V_\Psi$ then any $l + 1$ elements of the limit function $\mathbf{f_\Lambda} = (S_B)^\infty \Lambda$ are linearly dependent. In particular, if $l = 1$ then $\mathbf{f_\Lambda}(x) = f(x)v$ with $v \in V_\Psi$.

From now on, for the sake of simplicity, we assume $l = 1$.

In case the function vector $\Phi = (\phi_1, \phi_2, \ldots, \phi_m)^T$ is also refinable and satisfies the refinement equation

$$\Phi = \sum_{\alpha \in \mathbb{Z}^d} A_\alpha \, \Phi(2 \cdot - \alpha), \tag{2.5}$$

with $\mathbf{A} = (A_\alpha)_{\alpha \in \mathbb{Z}^d}$ a matrix sequence of real $n \times n$ matrices with compact support, it is not difficult to prove that the vector function $\Theta := \Phi * \Psi$ is again refinable satisfying

$$\Theta = \sum_{\gamma \in \mathbb{Z}^d} C_\gamma \, \Theta(2 \cdot - \gamma) \; ,$$

where $\mathbf{C} = (C_\gamma)_{\gamma \in \mathbb{Z}^d}$ is a matrix sequence of $(nm \times nm)$–matrices. To be able to describe $\mathbf{C}$ explicitly, we introduce the Kronecker type notion of convolving two matrix sequences $\mathbf{A}$ and $\mathbf{B}$, namely

$$(\mathbf{A} \circledast \mathbf{B})_\gamma = \sum_{\alpha \in \mathbb{Z}^d} A_\alpha \otimes B_{\gamma - \alpha} \; , \quad \gamma \in \mathbb{Z}^d \; , \tag{2.6}$$

the symbol $\otimes$ denoting the Kronecker product of matrices. Using this notion of matrix sequence convolution, it can be proved that [6]

$$\mathbf{C} = \frac{1}{2^d}(\mathbf{A} \circledast \mathbf{B}) . \tag{2.7}$$

Moreover, if S_A and S_B both generate convergent subdivision schemes, the subdivision scheme generated by the subdivision operator S_C associated with the refinement matrix mask $\mathbf{C}$, is again convergent (see [6] and [7] for details).

Now, suppose we want to represent a function $f_\Lambda = \sum_{\alpha \in \mathbb{Z}^d} (\Lambda_\alpha)^T \Theta(\cdot - \alpha)$ in S_Θ, with

$$S_\Theta := \text{span}\{\theta_1(\cdot - \alpha), \ldots, \theta_{nm}(\cdot - \alpha) ; \ \alpha \in \mathbb{Z}^d\} ,$$

and $\Lambda \in (\ell^\infty(\mathbb{Z}^d))^{nm}$. Because of the refinability of Θ, we know that

$$f_\Lambda = \sum_{\alpha \in \mathbb{Z}^d} (\Lambda_\alpha^{(k)})^T \Theta(2^k \cdot - \alpha) ,$$

with $\Lambda^{(k)}$ obtained by repeated application of the subdivision operator S_C starting with $\Lambda^{(0)} := \Lambda$. Then, assuming that the subdivision schemes associated with Φ and Ψ both converge, we can take any nonzero entries of the vector sequence $\Lambda^{(k)}$, for example $(\Lambda^{(k)})_1$, as an approximation of f_Λ at the dyadic points. On the other hand, since the support of the mask $\mathbf{C}$ is the convex hull of the union of those of $\mathbf{A}$ and $\mathbf{B}$ and the dimensions of each matrix C_α is the product of those of A_α and B_α, a strategy reducing the computational cost would be convenient. To this end we first need the following Lemma whose proof is trivial.

Lemma 2.2. *Any vector sequence* $\Lambda \in (\ell^\infty(\mathbb{Z}^d))^{nm}$ *can be written as*

$$\Lambda := \sum_{i=1}^m \tilde{\Lambda}^i \circledast \mathbf{E}^i , \quad \tilde{\Lambda}^i \in (\ell^\infty(\mathbb{Z}^d))^n, \quad \mathbf{E}^i \in (\ell^\infty(\mathbb{Z}^d))^m , \ i = 1, \ldots, m ,$$

where the n–dimensional vector sequences $\tilde{\Lambda}^i$, $i = 1, \ldots, m$ *are defined as*

$$\tilde{\Lambda}^i_\alpha := \begin{pmatrix} (\Lambda_\alpha)_i \\ (\Lambda_\alpha)_{i+m} \\ (\Lambda_\alpha)_{i+2m} \\ \vdots \\ (\Lambda_\alpha)_{i+(n-1)m} \end{pmatrix} , \quad \alpha \in \mathbb{Z}^d, \ i = 1, \ldots, m$$

and the *m–dimensional vector sequences* $\mathbf{E}^i$, $i = 1, \ldots, m$ are "δ-sequences", i.e.,

$$
\mathbf{E}^i_\alpha := \delta_{\alpha,0}
\begin{pmatrix}
0 \\
\vdots \\
0 \\
1 \\
0 \\
\vdots \\
0
\end{pmatrix}
\leftarrow i, \quad \alpha \in \mathbb{Z}^d, \quad i = 1, \ldots, m \ .
$$

With this factorization of $\boldsymbol{\Lambda}$, we are in a position to prove the theorem below.

Theorem 2.3. *Let $k \geq 1$, $\boldsymbol{\Lambda}^{(k)} = S_C \boldsymbol{\Lambda}^{(k-1)}$, and $(\tilde{\boldsymbol{\Lambda}}^i)^{(k)} = S_A (\tilde{\boldsymbol{\Lambda}}^i)^{(k-1)}$, $(\mathbf{E}^i)^{(k)} = S_B(\mathbf{E}^i)^{(k-1)}$, $i = 1, \ldots, m$, with $\boldsymbol{\Lambda}^{(0)} = \boldsymbol{\Lambda}$, $(\tilde{\boldsymbol{\Lambda}}^i)^{(0)} = \tilde{\boldsymbol{\Lambda}}^i$, $(\mathbf{E}^i)^{(0)} = \mathbf{E}^i$ and $\boldsymbol{\Lambda} = \sum_{i=1}^m \tilde{\boldsymbol{\Lambda}}^i \circledast \mathbf{E}^i$.*

Then

$$
\boldsymbol{\Lambda}^{(k)} = \frac{1}{2^{d \cdot k}} \sum_{i=1}^m (\tilde{\boldsymbol{\Lambda}}^i)^{(k)} \circledast (\mathbf{E}^i)^{(k)} \ . \tag{2.8}
$$

Proof. The proof is done by induction on k. For $k = 1$ the first step of the subdivision algorithm produces $\Lambda^{(1)}_\alpha = \sum_{\beta \in \mathbb{Z}^d} C^T_{\alpha - 2\beta} \Lambda_\beta$, for all $\alpha \in \mathbb{Z}^d$. Now, by (2.7) we know that

$$
C^T_{\alpha - 2\beta} = \frac{1}{2^d} \sum_{\gamma \in \mathbb{Z}^d} A^T_\gamma \otimes B^T_{\alpha - 2\beta - \gamma} \ ,
$$

implying

$$
\Lambda^{(1)}_\alpha = \frac{1}{2^d} \sum_{i=1}^m \Big(\sum_{\beta \in \mathbb{Z}^d} \Big(\sum_{\gamma \in \mathbb{Z}^d} A^T_\gamma \otimes B^T_{\alpha - 2\beta - \gamma} \sum_{\delta \in \mathbb{Z}^d} \tilde{\Lambda}^i_\delta \otimes E^i{}_{\beta - \delta} \Big) \Big), \text{ for all } \alpha \in \mathbb{Z}^d \ .
$$

Letting $\ell := \beta - \delta$ and $s := \alpha - 2\delta - \gamma$ and using the fact that for any four given matrices with suitable dimensions, $(M_1 \otimes M_2)(M_3 \otimes M_4) = (M_1 \, M_2) \otimes (M_3 \, M_4)$ holds, we can write

$$
\Lambda^{(1)}_\alpha = \frac{1}{2^d} \sum_{i=1}^m \Big(\sum_{s \in \mathbb{Z}^d} \Big(\sum_{\delta \in \mathbb{Z}^d} A^T_{\alpha - s - 2\delta} \tilde{\Lambda}^i_\delta \otimes \sum_{\ell \in \mathbb{Z}^d} B^T_{s - 2\ell} E^i{}_\ell \Big) \Big),
$$

$$
= \frac{1}{2^d} \sum_{i=1}^m \Big(\sum_{s \in \mathbb{Z}^d} (\tilde{\Lambda}^i)^{(1)}_{\alpha - s} \otimes (E^i)^{(1)}_s \Big), \text{ for all } \alpha \in \mathbb{Z}^d,
$$

thus proving (2.8) for $k = 1$. Now, assuming that (2.8) holds for $k = j$, $j \geq 1$ we have for all $\alpha \in \mathbf{Z}^d$

$$\Lambda_\alpha^{(j+1)}$$

$$= \sum_{i=1}^m \left(\sum_{\beta \in \mathbf{Z}^d} \frac{1}{2^d} \left(\sum_{\gamma \in \mathbf{Z}^d} A_\gamma^T \otimes B_{\alpha-2\beta-\gamma}^T \frac{1}{2^{d \cdot j}} \sum_{\delta \in \mathbf{Z}^d} (\tilde{\Lambda}^i)_\delta^{(j)} \otimes (E^i)_{\beta-\delta}^{(j)} \right) \right)$$

$$= \frac{1}{2^{d \cdot (j+1)}} \sum_{i=1}^m \left(\sum_{s \in \mathbf{Z}^d} (\tilde{\Lambda}^i)_{\alpha-s}^{(j+1)} \otimes (E^i)_s^{(j+1)} \right), \text{ for all } \alpha \in \mathbf{Z}^d,$$

thus concluding the proof. $\qquad\qquad\qquad\qquad\qquad\qquad\qquad\square$

In light of Theorem 2.3, any function $f_\Lambda \in S_\Theta$ with $\Lambda \in (\ell^\infty(\mathbf{Z}^d))^{nm}$ can be approximated at the dyadic points by adding the convolution of the s-calar sequences given by all the first components of $(\tilde{\Lambda}^i)^{(k)}$, $i = 1, \ldots, m$ and $(E^i)^{(k)}$, $i = 1, \ldots, m$, i.e.,

$$f_\Lambda\left(\frac{\alpha}{2^k}\right) \approx \frac{1}{2^{d \cdot k}} \sum_{i=1}^m \sum_{\beta \in \mathbf{Z}^d} ((\tilde{\Lambda}^i)_\beta^{(k)})_1 ((E^i)_{\alpha-\beta}^{(k)})_1, \quad \alpha \in \mathbf{Z}^d.$$

Remark 2.4. The sequences $(\tilde{\Lambda}^i)^{(k)}$, $i = 1, \ldots, m$ and $(E^i)^{(k)}$, $i = 1, \ldots, m$ can also be written as

$$(\tilde{\Lambda}^i)_\alpha^{(k)} = \sum_\beta (A_{\alpha-2^k\beta}^{(k)})(\tilde{\Lambda}^i)_\beta,$$

$$(E^i)_\alpha^{(k)} = \sum_\beta (B_{\alpha-2^k\beta}^{(k)})(E^i)_\beta,$$

where the iterated matrix sequence $\mathbf{D}^{(k)}$ of any given sequence $\mathbf{D}$ is defined as

$$\mathbf{D}^{(1)} := \mathbf{D}, \quad D_\alpha^{(k)} := \sum_{\beta \in \mathbf{Z}^d} D_\beta^{(k-1)} D_{\alpha-2\beta}, \quad \alpha \in \mathbf{Z}^d, \quad \text{for } k > 1.$$

It follows that the m vector sequences $(\tilde{\Lambda}^i)^{(k)}$, $i = 1, \ldots, m$ can be obtained by applying the iterated matrix sequence $\mathbf{A}^{(k)}$ to all different start vectors $\tilde{\Lambda}^i$, $i = 1, \ldots, m$. In particular, the vector sequences $(E^i)^{(k)}$, $i = 1, \ldots, m$ are just the columns of the matrix sequence $\mathbf{B}^{(k)}$.

Obviously, such a way of splitting the subdivision algorithm appears to be particularly suitable for parallel implementations. The analysis of the computational cost of a parallel implementation is currently under investigation.

3 A Subdivision–Like Scheme for the Non–Complete Refinable Case

The aim of this section is to discuss a subdivision–like scheme to represent any function f_Λ in the shift–invariant space

$$S_\Theta := \text{span}\{\theta_1(\cdot - \alpha), \dots, \theta_{nm}(\cdot - \alpha) \; ; \; \alpha \in \mathbb{Z}^d\}$$

with $\Lambda \in (\ell^\infty(\mathbb{Z}^d))^{nm}$, $\Theta := \Phi * \Psi$, under the assumption that only the second factor in the convolution product is refinable.

For this purpose, we prove the following theorem.

Theorem 3.1. *Let $f_\Lambda \in S_\Theta$ be such that $f_\Lambda = \sum_{\alpha \in \mathbb{Z}^d} (\Lambda_\alpha)^T \Theta(\cdot - \alpha)$ with $\Theta := \Phi * \Psi$, $\Lambda \in (\ell^\infty(\mathbb{Z}^d))^{nm}$, $\Lambda = \sum_{i=1}^m \tilde\Lambda^i \oplus \mathbf{E}^i$. Let Ψ be refinable with an associated convergent subdivision scheme defined by means of the subdivision operator S_B. Furthermore, let us set $g_{\tilde\Lambda^i} := \sum_{\alpha \in \mathbb{Z}^d} (\tilde\Lambda_\alpha^i)^T \Phi(\cdot - \alpha)$, $i = 1, \dots, m$ and let us consider for $k \geq 1$ the scalar sequences $(\mathbf{g}^i)^{(k)} := (g_{\tilde\Lambda^i}(\frac{\alpha}{2^k}))_{\alpha \in \mathbb{Z}^d}$, $(\mathbf{E}^i)_1^{(k)}$, $i = 1, \dots, m$ being $(\mathbf{E}^i)_1^{(k)}$ the first component of the vector sequence obtained by repeated application of S_B starting with $(\mathbf{E}^i)^{(0)} := \mathbf{E}^i$.*

Then it holds

$$\lim_{k \to \infty} \left\| \mathbf{f}_\Lambda^{(k)} - \sum_{i=1}^m (\mathbf{g}^i)^{(k)} * (\mathbf{E}^i)_1^{(k)} \right\|_\infty = 0 \; , \tag{3.1}$$

where $\mathbf{f}_\Lambda^{(k)} := (f_\Lambda(\frac{\alpha}{2^k}))_{\alpha \in \mathbb{Z}^d}$.

Proof. As $f_\Lambda \in S_\Theta$ with $\Lambda \in (\ell^\infty(\mathbb{Z}^d))^{nm}$, by Lemma 2.2 it follows that

$$\begin{aligned} f_\Lambda &= \sum_{\alpha \in \mathbb{Z}^d} (\Lambda_\alpha)^T \Theta(\cdot - \alpha) \\[2mm] &= \sum_{i=1}^m \left(\sum_{\alpha \in \mathbb{Z}^d} \sum_{\beta \in \mathbb{Z}^d} (\tilde\Lambda_\beta^i \otimes E^i{}_{\alpha - \beta})^T \Theta(\cdot - \alpha) \right) , \\[2mm] &= \sum_{i=1}^m \left(\sum_{\alpha \in \mathbb{Z}^d} (\tilde\Lambda_\alpha^i)^T \Phi(\cdot - \alpha) * \sum_{\alpha \in \mathbb{Z}^d} (E^i{}_\alpha)^T \Psi(\cdot - \alpha) \right) . \end{aligned} \tag{3.2}$$

Setting $g_{\tilde\Lambda^i} := \sum_{\alpha \in \mathbb{Z}^d} (\tilde\Lambda_\alpha^i)^T \Phi(\cdot - \alpha)$ and $h_{E^i} := \sum_{\alpha \in \mathbb{Z}^d} (E^i{}_\alpha)^T \Psi(\cdot - \alpha)$, $i = 1, \dots, m$, f_Λ can be written as

$$f_\Lambda = \sum_{i=1}^{m} g_{\tilde\Lambda^i} * h_{E^i} \,.$$

Now, because of the refinability of Ψ, each function h_{E^i} can be expressed as

$$h_{E^i} = \sum_{\alpha \in \mathbb{Z}^d} ((E^i)_\alpha^{(k)})^T \Psi(2^k \cdot -\alpha) \,, \quad i = 1, \dots, m \,,$$

so that applying a tensor product rectangular rule to each convolution integral $g_{\tilde\Lambda^i} * h_{E^i}$, it follows that for all i

$$\lim_{k\to\infty} \left\| (g_{\tilde\Lambda^i} * h_{E^i})\left(\frac{\alpha}{2^k}\right) - \frac{1}{2^{d\cdot k}}\left(\sum_{\beta \in \mathbb{Z}^d} g_{\tilde\Lambda^i}\left(\frac{\beta}{2^k}\right) h_{E^i}\left(\frac{\alpha - \beta}{2^k}\right) \right) \right\|_\infty = 0 \,.$$

In conclusion, setting $\mathbf{f}_\Lambda^{(k)} := (f_\Lambda(\frac{\alpha}{2^k}))_{\alpha \in \mathbb{Z}^d}$, (3.1) holds true. $\square$

Remark 3.2. Obviously, in case two given sequences, say $\mathbf{a} \in (\ell^\infty(\mathbb{Z}^d))^n$ and $\mathbf{b} \in (\ell^\infty(\mathbb{Z}^d))^m$, are such that $\Lambda = \mathbf{a} \circledast \mathbf{b}$, the subdivision–like scheme as well as the subdivision splitting discussed in Section 2 turn out to have a simplified structure.

4 Some Examples

To give an example of the application of convolution of function vectors and subdivision (or subdivision–like) algorithm for the representation of surfaces we refer to [15], where the author proposed a strategy to construct new families of B–splines on the three and four directional mesh of the plane, τ_3 and τ_4 respectively. Given a simple polygon P made of a small number of triangles, the author first studied the existence of so called simple splines with support P of minimal degree. Then he proceeded by defining composed B–splines by repeated convolution of simple splines with piecewise affine or quadratic box–splines. Now, the composed B–splines defined in [15] can be also constructed by means of subdivision (or subdivison–like) algorithms using a vector approach. Referring to τ_3 we take into account here the simplest splines over the triangle and the unit square (Δ and Σ polygons, see Fig. 1), i.e., their characteristic functions denoted by δ_{-1} and σ_{-1} (keeping the notation in [15]). Furthermore, we consider h_0, the piecewise affine pyramid over the H–polygon, also given in Fig. 1. We then convolve these functions setting $\Phi := \binom{\delta_{-1}}{\sigma_{-1}}$ and $\Psi := h_0$, thus obtaining the refinable function vector $\Theta = \binom{\theta_1}{\theta_2}$. Figure 2 shows the graphs of θ_1, θ_2 after $k = 3$ steps of applying the subdivision scheme. It should be noted that in case we do not use the vector approach (as done in [15]), we have to use a subdivision–like algorithm in order to represent θ_1 because of the non–refinability of the function δ_{-1}. Referring to τ_4 we consider two characteristic functions denoted by $\tilde{\sigma}_{-1}$ and λ_{-1} over the domains given in Fig. 3 and the two pyramidal hat functions, $\tilde{\sigma}_0$ and λ_0 respectively, over the same domains. Setting $\Phi := \binom{\tilde{\sigma}_{-1}}{\lambda_{-1}}$ and $\Psi := \binom{\tilde{\sigma}_0}{\lambda_0}$ we are not able to perform their convolution via a subdivision scheme directly, because the subdivision scheme associated with the refinement equation of the characteristic function λ_{-1} does not converge, and neither does the one associated with Φ. Obviously, in this case to compute $\Phi * \Psi$ a subdivision–like algorithm has to be used, where the subdivision–like algorithm is needed to represent the functions of the vector $\lambda_{-1} * \Psi$. A subdivision scheme can also be used in order to represent the functions of the vector $\tilde{\sigma}_{-1} * \Psi$. Figures 4 and 5 display the obtained functions with $k = 3$.

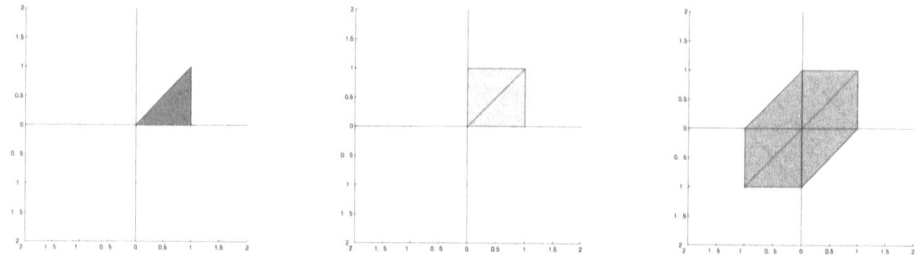

Fig. 1. Δ–*polygon,* Σ–*polygon and* H–*polygon*

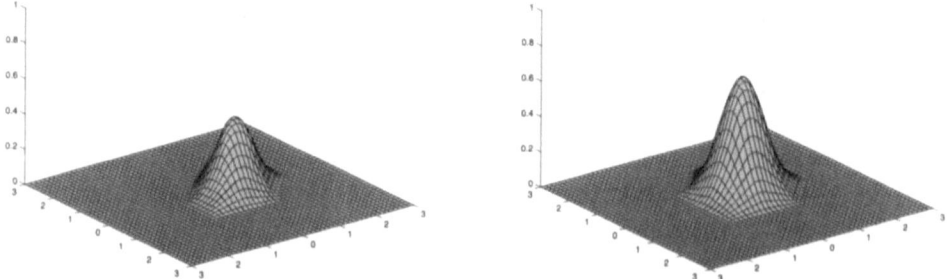

Fig. 2. $\delta_{-1} * h_0$ *and* $\sigma_{-1} * h_0$

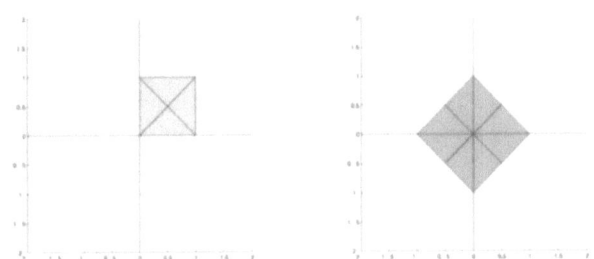

Fig. 3. Σ-polygon and Λ-polygon

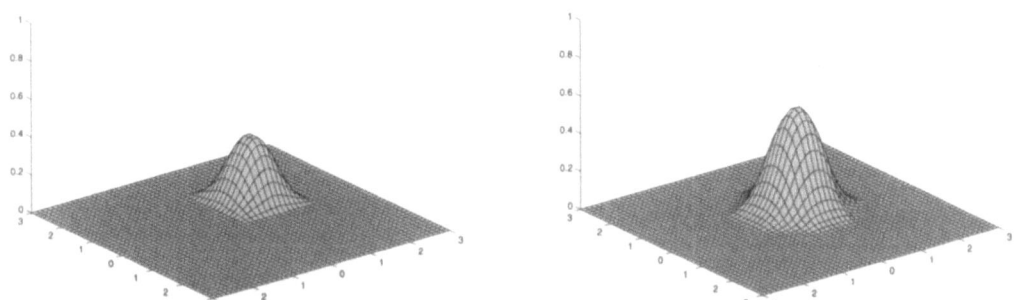

Fig. 4. $\tilde{\sigma}_{-1} * \tilde{\sigma}_0$ and $\tilde{\sigma}_{-1} * \lambda_0$

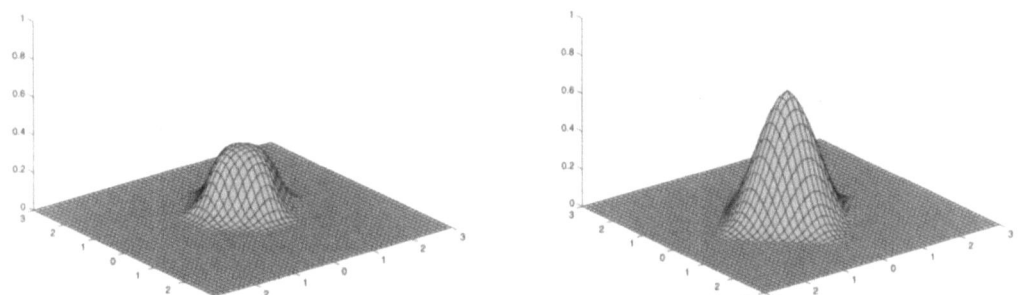

Fig. 5. $\lambda_{-1} * \tilde{\sigma}_0$ and $\lambda_{-1} * \lambda_0$

References

[1]   A. Cavaretta, W. Dahmen, C. A. Micchelli: *Stationary Subdivision*, Memoirs Amer. Math. Soc. **93**, no. 453, Providence, RI, 1991.

[2]   M. Charina, C. Conti, K. Jetter: *Image data compression with three-directional splines*, in: Proceedings of the SPIE conference 'Wavelets Application in Signal and Image Processing VIII', A. Aldroubi, A. F. Laine, M. Unser (eds.), San Diego, to appear (2000).

[3]   C. K. Chui: *Multivariate Splines*, CBMS–NSF Series Applied Math. **54**, Soc. Ind. Appl. Math., Philadelphia 1988.

[4]   A. Cohen, N. Dyn, D. Levin: *Stability and inter–dependence of matrix subdivision schemes*, in: Advanced Topics in Multivariate Approximation, F. Fontanella, K. Jetter, P. J. Laurent (eds.), World Scientific, Singapore 1996, pp. 33–45.

[5]   C. Conti, K. Jetter: *A new subdivision method for bivariate splines on the four–directional mesh*, J. Comp. Applied Math. **119** (2000), 81–96.

[6]   C. Conti, K. Jetter: *A note on convolving refinable function vectors*, in: Curve and Surface Fitting: Saint–Malo 1999, A. Cohen, C. Rabut, L. L. Schumaker (eds.), Vanderbilt University Press, Nashville 2000, pp. 135–142.

[7]   C. Conti: *Convergence of Convolved Vector Subdivision Schemes*, Technical Report, 2000.

[8]   S. Dahlke, W. Dahmen, V. Latour: *Smooth refinable functions and wavelets obtained by convolution products*, Appl. Comp. Harm. Anal. **2** (1995), 68–84.

[9]   W. Dahmen, C. A. Micchelli: *Biorthogonal wavelet expansion*, Contruct. Approx. **13** (1997), 293–328.

[10]   N. Dyn: *Subdivision schemes in computer–aided geometric design*, in: Advances in Numerical Analysis, Vol. II: Wavelets, Subdivision Algorithms and Radial Basis Functions, W. Light (ed.), Clarendon Press, Oxford 1992, pp. 36–104.

[11]   X. Guillet: *Interpolation by new families of B-splines on uniform meshes of the plane*, in: Surface Fitting and Multiresolution Methods, A. Le Méhauté, C. Rabut, L. L. Schumaker (eds.), World Scientific, Singapore 1997, pp. 183–190.

[12]   R. Q. Jia, S. Riemenschneider, D. X. Zhou: *Vector subdivision schemes and multiple wavelets*, Math. Comp. **67** (1998), 1533–1563.

[13]   C. A. Micchelli, T. Sauer: *On Vector Subdivision*, Math. Z., to appear.

[14]   S. D. Riemenschneider, Z.-W. Shen, *Box splines, cardinal series, and wavelets*, in: Approximation Theory and Functional Analysis, C. K. Chui (ed.), Academic Press, San Diego 1991, pp. 133–149.

[15]   P. Sablonnière: *B-splines on uniform meshes of the plane*, in: Advanced Topics in Multivariate Approximation, F. Fontanella, K. Jetter, P. J. Laurent (eds.), World Scientific, Singapore 1996, pp. 323–337.

Address:

Costanza Conti
Dipartimento di Energetica
Università di Firenze
Via Lombroso 6 / 17
I–50134 Firenze
Italy

International Series of Numerical Mathematics
Vol. 137, ©2001 Birkhäuser Verlag Basel/Switzerland

Best One–Sided L^1–Approximation by $B^{2,1}$–Blending Functions

Dimiter Dryanov, Werner Haußmann and Petar Petrov

Abstract

Let $f \in C^{2,1}\left([-1,1]^2\right)$ be a real function satisfying $\partial^3 f / \partial x^2 \partial y \geq 0$ on $[-1,1]^2$. We prove existence and uniqueness of the best one–sided L^1–approximant h^* to f from above (resp. h_* from below) with respect to the *infinite dimensional linear space* of all $B^{2,1}$–blending functions:

$$B^{2,1} := \left\{ g \in C^{2,1}\left([-1,1]^2\right) : \frac{\partial^3 g}{\partial x^2 \partial y} = 0 \right\}.$$

The unique best one–sided approximants h^* (resp. h_*) are characterized by Hermite type interpolation conditions with respect to canonical point sets $V^* := \{(x,y) \in [-1,1]^2 : y = 2|x| - 1\}$ resp. $\Lambda_* := \{(x,y) \in [-1,1]^2 : y = -2|x| + 1\}$.

1 Preliminaries

The classical univariate algebraic polynomials have a natural multivariate counterpart, the so called *blending functions*. The real linear space $B^{m,n}\left(I^2\right)$ of all blending functions of order (m,n) is defined by

$$B^{m,n}\left(I^2\right) := \left\{ h \in C^{m,n}\left(I^2\right) : D^{m,n} h := \frac{\partial^{m+n} h}{\partial x^m \, \partial y^n} = 0 \text{ on } I^2 \right\},$$

where $I := [-1,1]$ and

$$C^{m,n}\left(I^2\right) := \left\{ f : I^2 \to \mathbb{R} : D^{i,j} f \in C\left(I^2\right), \ 0 \leq i \leq m, \ 0 \leq j \leq n \right\}.$$

The elements of the linear space of blending functions $B^{m,n} := B^{m,n}\left(I^2\right)$ are usually called $B^{m,n}$–blending functions. $B^{m,n}$ is an infinite dimensional vector space while the univariate analogue, the space of all algebraic polynomials of degree less than or equal to $m - 1$,

$$\pi_m := \left\{ p \in C^m\left(I\right) : \frac{d^m p}{d x^m} = 0 \right\},$$

is of dimension $m < \infty$.

Any blending function $h \in B^{m,n}$ can be represented in the form

$$h(x,y) = \sum_{k=0}^{m-1} a_k(y)x^k + \sum_{l=0}^{n-1} b_l(x)y^l, \qquad a_k \in C^n(I), \, b_l \in C^m(I).$$

Note that the above representation is not unique. Conversely, each function of this form is a $B^{m,n}$–blending function, so

$$B^{m,n} = \left\{ h \in C^{m,n}(I^2) \; : \; h(x,y) = \sum_{k=0}^{m-1} a_k(y)x^k + \sum_{l=0}^{n-1} b_l(x)y^l \right\}.$$

Despite the fact that $B^{m,n}$ is infinite–dimensional, some of the approximation results for univariate algebraic polynomials have their natural counterparts in the theory of approximation by blending functions. As an example in the case of unconstrained best L^1–approximation by algebraic polynomials we mention *the classical theorem of A. Markov* (see Achieser [1, pp. 82–85] or DeVore and Lorentz [10, p. 87]) and the corresponding result of Haußmann and Zeller [12] for best L^1–approximation by blending functions on I^2, where *the canonical point set* forms a so–called *blending grid* consisting of appropriate horizontal and vertical lines. If the space of approximating functions consists of the infinite dimensional space of harmonic functions on the multidimensional unit ball, then we have a concentric sphere as a canonical point set (see [3] for details). For further references on *canonical point set* characterizations of best multidimensional L^1–approximation we refer to the survey paper [13].

Concerning multivariate best one–sided L^1–approximation, it has been shown in [6] that blending grids of horizontal and vertical lines occur as canonical point sets in the case of *one–sided L^1–approximation by blending functions*, but only under some additional restrictions. Best one–sided L^1–approximation by harmonic functions has been studied in [4]. A general approach to questions of existence, uniqueness and characterization of best one–sided L^1–approximation is given in [2]. In a recent paper [11] we prove that for functions f, satisfying $D^{1,1}f \geq 0$ on I^2, the unique best one–sided L^1–approximant by $B^{1,1}$–blending functions is characterized in terms of *Hermite interpolation* with respect to diagonals of I^2 that play the role of canonical point sets in this case. *The occurrence of the diagonals of I^2 as canonical point sets in the case of approximation by blending functions is a phenomenon which to the best of our knowledge has not been observed before.*

In the present paper we consider real valued functions $f \in C^{2,1}(I^2)$ satisfying $\partial^3 f/\partial x^2 \partial y \geq 0$ on I^2. We characterize the unique best one–sided L^1–

approximants h^* to f from above resp. h_* from below with respect to the infinite dimensional space $B^{2,1}$. The characterization is given by *Hermite interpolation on the canonical point set*

$$V^* := \left\{ (x,y) \in I^2 \; : \; y = 2|x| - 1 \right\}$$

when we approximate from above resp. *on the canonical set*

$$\Lambda_* := \left\{ (x,y) \in I^2 \; : \; y = -2|x| + 1 \right\}$$

in the case of approximation from below. Note that the Hermite interpolation on a fixed grid is a linear operator, while best one–sided L^1–approximation is not linear in general.

The $B^{m,n}$–blending functions are natural generalizations of univariate polynomials to the bivariate setting. So, the main results in this article (Theorem 1 and Theorem 2) can be considered as bivariate blending–type extensions of theorems by Bojanić and DeVore (see [5] and [9] for details) concerning univariate best one–sided L^1–approximation by Chebyshev systems (in particular algebraic polynomials). A convenient introduction to the topic of best one–sided L^1–approximation can be found in the book by Pinkus [14, Chapter 5]. The previous work in this area has mainly dealt with the case of finite–dimensional spaces of approximating functions.

2 Main Results

A function $h^* \in B^{2,1}$ is defined to be a best one–sided L^1–approximant from above to $f \in C(I^2)$ with respect to $B^{2,1}(I^2)$ if $h^* - f \geq 0$ on I^2 and

$$\int_{I^2} (h^* - f) \leq \int_{I^2} (h - f) \quad \text{for all } h \in B^{2,1} \text{ such that } h - f \geq 0 \text{ on } I^2.$$

Analogously, we define a *best one–sided L^1–approximant from below* and we denote it by h_*.

Our main results concerning best one–sided L^1–approximation by $B^{2,1}$–blending functions are the next two theorems.

Theorem 1. *Let $f \in C^{2,1}(I^2)$ satisfy $D^{2,1}f \geq 0$ on I^2. Then f possesses a unique best one–sided L^1–approximant h^* from above with respect to $B^{2,1}$.*

The unique best one–sided L^1–approximant h^* is characterized by the following Hermite interpolation conditions

$$h^*_{|V^*} = f_{|V^*} \quad and \quad \operatorname{grad} h^*_{|V^*} = \operatorname{grad} f_{|V^*} \qquad \left(h^* \in B^{2,1}\right). \qquad (2.1)$$

Theorem 2. Let $f \in C^{2,1}(I^2)$ satisfy $D^{2,1}f \geq 0$ on I^2. Then f possesses a unique best one–sided L^1–approximant h_* from below with respect to $B^{2,1}$. The unique best one–sided L^1–approximant h_* is characterized by the following Hermite interpolation conditions

$$h_*{}_{|\Lambda_*} = f_{|\Lambda_*} \quad and \quad \operatorname{grad} h_*{}_{|\Lambda_*} = \operatorname{grad} f_{|\Lambda_*} \qquad \left(h_* \in B^{2,1}\right). \qquad (2.2)$$

The proofs of Theorems 1 and 2 (see Section 6) are based on Hermite interpolation results with respect to infinite sets which will be given in Theorems 3 and 4.

3 Hermite Interpolation by $B^{2,1}$–Blending Functions

In this section we shall prove that a given function f and its gradient grad f can be simultaneously uniquely interpolated on the canonical point sets V^* resp. Λ_* with respect to $B^{2,1}$–blending functions on I^2.

Theorem 3. For $f \in C^{2,1}(I^2)$ define

$$F^*(t) := \int_{-1}^{t} \frac{1}{v+1} \left[f^{(0,1)}\left(\frac{v+1}{2}, v\right) - f^{(0,1)}\left(-\frac{v+1}{2}, v\right) \right] dv.$$

Then the $B^{2,1}$–blending function

$$h^*(x,y) := x F^*(y) - \int_0^x F^*(2|t| - 1) \, dt + \int_0^x f^{(1,0)}(t, 2|t| - 1) \, dt \qquad (3.1)$$

$$+ \frac{1}{2} \int_{-1}^y \left[f^{(0,1)}\left(\frac{t+1}{2}, t\right) + f^{(0,1)}\left(-\frac{t+1}{2}, t\right) \right] dt + f(0, -1)$$

is the unique $B^{2,1}$–blending interpolant to f which satisfies the interpolation conditions (2.1) on the point set V^*.

Proof. (i) First we prove that $h^* \in B^{2,1}(I^2)$. Clearly, $h^* \in C^{2,1}(I_-^2)$ and $h^* \in C^{2,1}(I_+^2)$, where $I_-^2 := \{(x,y) \in I^2 : x \in [-1,0]\}$ and $I_+^2 := \{(x,y) \in I^2 : x \in [0,1]\}$. Hence we have $h^* \in B^{2,1}(I_-^2)$ and $h^* \in B^{2,1}(I_+^2)$.

We need to verify that for any $y \in [-1,1]$

$$\lim_{x \to 0+} D^{i,j} h^*(x,y) = \lim_{x \to 0-} D^{i,j} h^*(x,y) \quad \text{for} \quad i = 0,1,2; \ j = 0,1,$$

which then implies that $h^* \in C^{2,1}(I^2)$; hence $h^* \in B^{2,1}(I^2)$. We are going to show only that $h^* \in C^{2,0}(I^2)$. For the other values of i and j the above equality follows along similar patterns. In order to show that $D^{2,0}h^* \in C(I^2)$ we calculate ($x \neq 0$):

$$D^{1,0}h^*(x,y) \ = \ F^*(y) - F^*(2|x|-1) + f^{(1,0)}(x,2|x|-1) \ ;$$

$$D^{2,0}h^*(x,y) \ = \ \frac{1}{x}\left[f^{(0,1)}(-|x|,2|x|-1) - f^{(0,1)}(|x|,2|x|-1)\right]$$
$$+ f^{(2,0)}(x,2|x|-1) + 2\,\text{sign}(x)\,f^{(1,1)}(x,2|x|-1) \ .$$

Taking the limit $x \to 0$ we obtain (since $-|x| < \xi_x < |x|$)

$$\lim_{x \to 0; x \neq 0} D^{2,0}h^*(x,y) \ = \ \lim_{x \to 0; x \neq 0}\left[-2\,\text{sign}(x)\,f^{(1,1)}(\xi_x,2|x|-1)\right.$$
$$\left. + f^{(2,0)}(x,2|x|-1) + 2\,\text{sign}(x)\,f^{(1,1)}(x,2|x|-1)\right]$$
$$= \ f^{(2,0)}(0,-1) \ .$$

It follows that $h^* \in C^{2,1}(I^2)$, thus $h^* \in B^{2,1}(I^2)$.

Suppose that for an integer $r \geq 3$ a function $f \in C^{i,j}(I^2)$, $i + j \leq r$ is given. Let $I_\varepsilon^2 := \{(x,y) \in (-\varepsilon,\varepsilon) \times I\}$, $\varepsilon \in (0,1]$. Straight calculations show that h^* does not belong to $C^{3,0}(I_\varepsilon^2)$ unless $f^{(2,1)}(0,-1) = 0$. Thus, h^* possesses no better smoothness than $C^{2,1}$ in general, no matter how smooth is the interpolated function f.

(ii) Next we show that h^* satisfies the interpolation conditions (2.1). Let $(x,y) \in V^*$, i.e. $y = 2|x| - 1$, $x \in [-1,1]$. We have

$$h^*(x,2|x|-1) \ = \ xF^*(2|x|-1) - \int_0^x F^*(2|t|-1)\,dt + \int_0^x f^{(1,0)}(t,2|t|-1)\,dt$$
$$+ \frac{1}{2}\int_{-1}^{2|x|-1}\left[f^{(0,1)}\left(\frac{t+1}{2},t\right) + f^{(0,1)}\left(-\frac{t+1}{2},t\right)\right]dt$$
$$+ f(0,-1)$$

$$= \quad x F^*(2|x| - 1) - \int_0^x F^*(2|t| - 1)\, dt$$

$$- \operatorname{sign}(x) \int_{-1}^{2|x|-1} \frac{t+1}{2} \frac{d}{dt} F^*(t)\, dt + \int_0^x f^{(1,0)}(t, 2|t| - 1)\, dt$$

$$+ \int_{-1}^{2|x|-1} f^{(0,1)} \left(\operatorname{sign}(x) \frac{t+1}{2}, t \right)\, dt + f(0, -1)$$

$$= \quad x F^*(2|x| - 1) - \int_0^x \left[F^*(2|t| - 1) + 2\operatorname{sign}(t)\, t \frac{d}{dt} F^*(2|t| - 1) \right] dt$$

$$+ \int_0^x \left[f^{(1,0)}(t, 2|t| - 1) + 2\operatorname{sign}(t)\, f^{(0,1)}(t, 2|t| - 1) \right] dt + f(0, -1)$$

$$= \quad x F^*(2|x| - 1) - \int_0^x \frac{d}{dt} \left[t F^*(2|t| - 1) \right] dt$$

$$+ \int_0^x \frac{d}{dt} f(t, 2|t| - 1)\, dt + f(0, -1)$$

$$= \quad f(x, 2|x| - 1) \ .$$

This shows that h^* and f match on V^*. Further we have

$$D^{1,0} h^*(x, 2|x| - 1) \quad = \quad F^*(2|x| - 1) - F^*(2|x| - 1) + f^{(1,0)}(x, 2|x| - 1)$$
$$= \quad f^{(1,0)}(x, 2|x| - 1)$$

and finally, we get

$$D^{0,1} h^*(x, 2|x| - 1) \quad = \quad x \frac{d}{dy} F^*(y)|_{y=2|x|-1} + \frac{1}{2} \Big[f^{(0,1)}(|x|, 2|x| - 1)$$

$$+ f^{(0,1)}(-|x|, 2|x| - 1) \Big]$$

$$= \quad \frac{\operatorname{sign}(x)}{2} \Big[f^{(0,1)}(|x|, 2|x| - 1) - f^{(0,1)}(-|x|, 2|x| - 1) \Big]$$

$$+ \frac{1}{2} \Big[f^{(0,1)}(|x|, 2|x| - 1) + f^{(0,1)}(-|x|, 2|x| - 1) \Big]$$

$$= \quad f^{(0,1)}(x, 2|x| - 1) \ .$$

Hence, $h^*|_{V^*} = f|_{V^*}$ and $\operatorname{grad} h^*|_{V^*} = \operatorname{grad} f|_{V^*}$. This shows the interpolating properties of h^*.

The fact that h^* is the *unique* $B^{2,1}$–blending interpolant to f satisfying (2.1) will be shown in Section 4.

Next we prove

Theorem 4. *Let $f \in C^{2,1}(I^2)$ be given. The function*

$$
\begin{aligned}
h_*(x,y) \;=\;& f(x, -2|x| + 1) \qquad\qquad\qquad\qquad\qquad\qquad\qquad (3.2)\\
& + x \int_{-2|x|+1}^{y} \frac{1}{1-t} \left[f^{(0,1)}\left(\frac{1-t}{2}, t\right) - f^{(0,1)}\left(\frac{t-1}{2}, t\right) \right] dt \\
& + \frac{1}{2} \int_{-2|x|+1}^{y} \left[f^{(0,1)}\left(\frac{1-t}{2}, t\right) + f^{(0,1)}\left(\frac{t-1}{2}, t\right) \right] dt
\end{aligned}
$$

is the unique $B^{2,1}$–blending function which satisfies the interpolation conditions (2.2) on the point set $\Lambda_* = \{(x,y) \in I^2 : y = -2|x| + 1\}$.

While in the proof of Theorem 3 it is verified that a given function h^* satisfies the interpolation conditions for the function f and its gradient, for Theorem 4 we present a proof which gives a construction of the interpolant. The construction is based on a former result (see [11]) concerning Hermite interpolation by $B^{1,1}$–blending functions on I^2. The representation of the interpolant h_* in (3.2) is somewhat different from that one given for h^* in (3.1), but the representation from (3.2) can be converted to a form corresponding to (3.1) and vice versa.

Proof of Theorem 4. We say that the functions $f, g \in C^{1,1}(I^2)$ are in equivalence relation with respect to a point set $S \subset I^2$ (denoted by $f \sim_S g$) if we have

$$
(f - g)_{|S} = 0 \quad \text{and} \quad \mathrm{grad}\,(f - g)_{|S} = 0.
$$

Let us consider (see [11] for details) the unique $B^{1,1}$ blending interpolant to $f \in C^{1,1}(I^2)$ on the main diagonal $\Delta := \{(x,y) \in I^2 \,|\, y = x\}$ of I^2:

$$
f(x,y) \sim_\Delta f(1,1) + \int_1^x f^{(1,0)}(t,t)\,dt + \int_1^y f^{(0,1)}(t,t)\,dt.
$$

Applying this formula to $f\left((\tilde{x} - 1)/2, y\right)$, $(\tilde{x}, y) \in I^2$ and then substituting $\tilde{x} = 2x + 1$, $x \in [-1, 0]$, we obtain the following equivalence relation

$$
f(x,y) \sim_{\Lambda_-} f(0,1) + \frac{1}{2} \int_1^{2x+1} f^{(1,0)}\left(\frac{t-1}{2}, t\right) dt + \int_1^y f^{(0,1)}\left(\frac{t-1}{2}, t\right) dt.
$$

on $\Lambda_- := \{(x,y) \in [-1,0] \times I \,|\, y = 2x + 1\}$.
Similarly, on $\Lambda_+ := \{(x,y) \in [0,1] \times I \,|\, y = -2x + 1\}$ we have

$$
f(x,y) \sim_{\Lambda_+} f(0,1) - \frac{1}{2} \int_1^{-2x+1} f^{(1,0)}\left(\frac{1-t}{2}, t\right) dt + \int_1^y f^{(0,1)}\left(\frac{1-t}{2}, t\right) dt.
$$

Suppose that $h_* = b_1(y)x + b_0(y) + a_0(x)$ is a $B^{2,1}$–blending function which satisfies (2.2) on Λ_*. It means that $f(x,y) - b_1(y)x \sim_{\Lambda_*} b_0(y) + a_0(x)$ on Λ_* and therefore by the above two formulas we get

$$f(x,y) - b_1(y)x \sim_{\Lambda_-} f(0,1) + \frac{1}{2}\int_1^{2x+1}\left[f^{(1,0)}\left(\frac{t-1}{2},t\right) - b_1(t)\right]dt \quad (3.3)$$
$$+ \int_1^y\left[f^{(0,1)}\left(\frac{t-1}{2},t\right) - b_1'(t)\frac{t-1}{2}\right]dt$$

on Λ_-, and

$$f(x,y) - b_1(y)x \sim_{\Lambda_+} f(0,1) - \frac{1}{2}\int_1^{-2x+1}\left[f^{(1,0)}\left(\frac{1-t}{2},t\right) - b_1(t)\right]dt \quad (3.4)$$
$$+ \int_1^y\left[f^{(0,1)}\left(\frac{1-t}{2},t\right) - b_1'(t)\frac{1-t}{2}\right]dt$$

on Λ_+. Hence, the uniqueness of the $B^{1,1}$–blending interpolant on the diagonals of I^2 (see [11] for details) implies that all $B^{2,1}$–blending interpolants to f on Λ_* are uniquely determined by the formulae (3.3) and (3.4) up to a $C^1(I)$–function $b_1(y)$. In view of the fact that the $(0,1)$–derivatives of the right hand sides of (3.3) and (3.4) have to agree on $\{(0,y) : y \in I\}$ we obtain:

$$b_1'(t)(1-t) = f^{(0,1)}\left(\frac{1-t}{2},t\right) - f^{(0,1)}\left(-\frac{1-t}{2},t\right).$$

Therefore

$$b_1(t) = F_*(t) + \text{const.}, \quad (3.5)$$

where

$$F_*(t) := \int_1^t \frac{1}{1-u}\left[f^{(0,1)}\left(\frac{1-u}{2},u\right) - f^{(0,1)}\left(-\frac{1-u}{2},u\right)\right]du.$$

Substituting $b_1(t)$ and making use of (3.5) in (3.3) and (3.4), we get

$$f(x,y) \sim_{\Lambda_-} f(0,1) + \frac{1}{2}\int_1^{2x+1}\left[f^{(1,0)}\left(\frac{t-1}{2},t\right) - F_*(t)\right]dt \quad (3.6)$$
$$+ \frac{1}{2}\int_1^y\left[f^{(0,1)}\left(\frac{1-t}{2},t\right) + f^{(0,1)}\left(\frac{t-1}{2},t\right)\right]dt + F_*(y)x$$

on Λ_-, and

$$f(x,y) \sim_{\Lambda_+} f(0,1) - \frac{1}{2}\int_1^{-2x+1}\left[f^{(1,0)}\left(\frac{1-t}{2},t\right) - F_*(t)\right]dt \quad (3.7)$$
$$+ \frac{1}{2}\int_1^y\left[f^{(0,1)}\left(\frac{1-t}{2},t\right) + f^{(0,1)}\left(\frac{t-1}{2},t\right)\right]dt + F_*(y)x$$

on Λ_+. So, the solution h_* of the interpolation problem (2.2) in $B^{2,1}(I^2)$ is uniquely determined by the formula

$$h_*(x,y) = f(0,1) - \frac{\operatorname{sign}(x)}{2} \int\limits_1^{-2|x|+1} \left[f^{(1,0)}\left(\operatorname{sign}(x)\frac{1-t}{2},t\right) - F_*(t) \right] dt \quad (3.8)$$

$$+ \frac{1}{2} \int\limits_1^y \left[f^{(0,1)}\left(\frac{1-t}{2},t\right) + f^{(0,1)}\left(\frac{t-1}{2},t\right) \right] dt + F_*(y)\, x \, .$$

In order to get the formula (3.2) we integrate by parts

$$-\frac{\operatorname{sign}(x)}{2} \int\limits_1^{-2|x|+1} \left[f^{(1,0)}\left(\operatorname{sign}(x)\frac{1-t}{2},t\right) - F_*(t) \right] dt$$

$$= -\frac{\operatorname{sign}(x)}{2} \int\limits_1^{-2|x|+1} f^{(1,0)}\left(\operatorname{sign}(x)\frac{1-t}{2},t\right) dt - x\, F_*(-2|x|+1)$$

$$+ \frac{\operatorname{sign}(x)}{2} \int\limits_1^{-2|x|+1} \left[f^{(0,1)}\left(\frac{1-t}{2},t\right) - f^{(0,1)}\left(\frac{t-1}{2},t\right) \right] dt$$

$$= -x\, F_*(-2|x|+1) + \int\limits_1^{-2|x|+1} \frac{d}{dt}[f\left(\operatorname{sign}(x)\frac{1-t}{2},t\right) dt$$

$$- \frac{1}{2} \int\limits_1^{-2|x|+1} \left[f^{(0,1)}\left(\frac{1-t}{2},t\right) + f^{(0,1)}\left(\frac{t-1}{2},t\right) \right] dt$$

$$= -x\, F_*(-2|x|+1) + f(x,-2|x|+1) - f(0,1)$$

$$- \frac{1}{2} \int\limits_1^{-2|x|+1} \left[f^{(0,1)}\left(\frac{1-t}{2},t\right) + f^{(0,1)}\left(\frac{t-1}{2},t\right) \right] dt \, .$$

Putting the above expression in (3.8) we obtain formula (3.2). Similar calculations as in the part (i) of the proof of Theorem 3 show that $h_* \in C^{2,1}(I^2)$. In particular, $\lim_{x\to 0, x\neq 0} D^{2,0}h_*(x,y) = f^{(2,0)}(0,1)$. On the other hand h_* does not belong to $C^{3,0}(I^2)$ unless $f^{(2,1)}(0,1) = 0$.

Remark 1. By the substitution $\text{sign}(x)(1-t)/2 = v$ in the first integral of (3.8) we obtain a formula similar to that of (3.1) for h_*, namely

$$
\begin{aligned}
h_*(x,y) &= xF_*(y) - \int_0^x F_*(-2\,|t|+1)\,dt + \int_0^x f^{(1,0)}(t,-2\,|t|+1)\,dt \\
&+ \frac{1}{2}\int_1^y \left[f^{(0,1)}\left(\frac{1-t}{2},t\right) + f^{(0,1)}\left(-\frac{1-t}{2},t\right)\right] dt + f(0,1)\,.
\end{aligned}
$$

Conversely, the blending interpolant h^* can be represented by

$$
\begin{aligned}
h^*(x,y) &= f(x,2|x|-1) \\
&+ x\int_{2|x|-1}^y \frac{1}{1+t}\left[f^{(0,1)}\left(\frac{1+t}{2},t\right) - f^{(0,1)}\left(-\frac{1+t}{2},t\right)\right] dt \\
&+ \frac{1}{2}\int_{2|x|-1}^y \left[f^{(0,1)}\left(\frac{1+t}{2},t\right) + f^{(0,1)}\left(-\frac{1+t}{2},t\right)\right] dt
\end{aligned}
$$

which is similar to (3.2). If the interpolant $h^* := h^*_{f(u,v)}$ of $f(x,y)$ satisfies the interpolation conditions (2.1), then it is easily seen that the $B^{2,1}$–blending function $h_*(x,y) = h^*_{f(u,-v)}(x,-y)$ satisfies the interpolation conditions (2.2). This means that we may construct h_* on the basis of an explicit form of h^* and conversely.

Remark 2. Note that the smoothness of the $B^{2,1}$–blending interpolants h^* and h_* does not increase with the smoothness of the interpolated function f. In general, h^* and h_* are only $C^{2,1}$–functions on I^2, while on I_+^2 and I_-^2 they have the same smoothness as f. For example, if $f^{(2,1)}(0,-1) \neq 0 \ (> 0)$, then h^* does not belong to $C^{(3,0)}\,(I^2)$; if $f^{(2,1)}(0,1) \neq 0 \ (> 0)$, then h_* is not a $C^{(3,0)}\,(I^2)$–function. Hence, if $D^{2,1}f > 0$ on I^2, then the corresponding one–sided best approximants h^* and h_* are $C^{2,1}$ and they have no higher smoothness *no matter how smooth is the approximated function f*.

Remark 3. If in Theorems 3 and 4 the function $f(x,y)$ is even with respect to x, then the $B^{2,1}$–blending interpolants h^* and h_* are also even with respect to x; hence h^* and h_* are $B^{1,1}$–blending functions in this particular case.

Remark 4. The representation (3.2) could be seen as a *Taylor type expansion up to "the first derivative"* around the one–dimensional set Λ_*. We like to mention that h^* is given by (3.1) in terms of $f(0,-1)$ and *the traces (interpolation conditions)* $D^{1,0}f_{|V_*}$ and $D^{0,1}f_{|V_*}$ of f. On the other hand h_* is given by (3.2) only in terms of the traces $f_{|\Lambda_*}$ and $D^{0,1}f_{|\Lambda_*}$ of the function f.

Example 1. Consider the polynomial $f_0(x,y) = x^2y$ on I^2. By (3.1) we have

$$
h^*(x,y) = (4/3)x^2|x| - x^2 + (1/12)(1+y)^3\,.
$$

Following Remark 1 we obtain a formula for h_* on the basis of the knowledge of h^*:

$$h_*(x,y) = -(4/3)x^2|x| + x^2 - (1/12)(1-y)^3 .$$

On the other hand $D^{3,0}h^*(x,y) = 8\,\mathrm{sign}(x)$, $x \neq 0$ and h^* does not belong to $C^{3,0}(I^2)$ (see Remark 2). $f_0(x,y)$ is an even function with respect to x, so $h^* \in B^{1,1}(I^2)$ (see Remark 3). We have $D^{2,1}f_0 = 2 > 0$ and Theorems 1 and 2 imply that h^* resp. h_* are the unique best one–sided L^1–approximants to f_0 from above resp. from below with respect to $B^{2,1}$. Hence, the best one–sided L^1–approximants to the polynomial f_0 are not polynomials, moreover they have limited $C^{2,1}$–smoothness (one may compare with the corresponding one–dimensional results [14] and with the unconstrained best L^1–approximation by blending functions [12]).

4   Error Representations for $f - h^*$ and $f - h_*$

First we shall prove error representations for the $B^{2,1}$–blending interpolants h^* and h_* in the form of Hermite–Lagrange.

Theorem 5.   *Let $f \in C^{2,1}(I^2)$. Then for the $B^{2,1}$–blending interpolants h^* and h_* from Theorems 3 and 4 the following error formulae hold true:*

$$\text{(a)} \quad f(x,y) - h^*(x,y) = -\frac{D^{2,1}f(\xi^*,\eta^*)}{24}(y - 2|x| + 1)^2(y + 4|x| + 1), \quad (4.1)$$

where $(\xi^*,\,\eta^*) = (\xi^*(x,y),\,\eta^*(x,y))$ belongs to I^2.

$$\text{(b)} \quad f(x,y) - h_*(x,y) = +\frac{D^{2,1}f(\xi_*,\eta_*)}{24}(y + 2|x| - 1)^2(-y + 4|x| + 1), \quad (4.2)$$

where $(\xi_*,\,\eta_*) = (\xi_*(x,y),\,\eta_*(x,y))$ belongs to I^2.

Proof. We shall prove the assertion for h^* only, since the proof for h_* is similar. Consider the auxiliary function $\varphi^*(x,y) := f(x,y) - h^*(x,y) + c \cdot r^*(x,y)$, where the $C^{2,1}(I^2)$–function $r^*(x,y) = \frac{1}{24}(y - 2|x| + 1)^2(y + 4|x| + 1)$ satisfies $D^{2,1}r^* = -1$ and $r^* \sim_{V^*} 0$ on the grid V^*. Let $(x_0, y_0) \in I^2 \setminus V^*$ and choose $c = c(x_0, y_0)$ such that $\varphi^*(x_0, y_0) = 0$. By Rolle's theorem there exists η^* in the open segment connecting y_0 and $2|x_0| - 1$ such that $D^{0,1}\varphi^*(x_0, \eta^*) = 0$. We have $\varphi^* \sim_{V^*} 0$ on V^*. Hence, from

$$D^{0,1}\varphi^*\left(-\frac{\eta^*+1}{2}, \eta^*\right) = D^{0,1}\varphi^*(x_0, \eta^*) = D^{0,1}\varphi^*\left(\frac{\eta^*+1}{2}, \eta^*\right) = 0$$

there exists $\xi^* \in (-1, 1)$ such that

$$D^{2,1}\varphi^*(\xi^*, \eta^*) = D^{2,1}f(\xi^*, \eta^*) + c(x_0, y_0) \cdot (-1) = 0,$$

and from here we conclude (4.1).

Completion of the proof of Theorem 3: uniqueness of h^*. Now we will show uniqueness of the Hermite interpolant h^*. For a given function $f \in C^{2,1}(I^2)$ let us suppose that there are two Hermite $B^{2,1}$–interpolants h^1 and h^2 satisfying the interpolation conditions (2.1). Hence, $h^0 := h^1 - h^2$ is a $B^{2,1}$–blending function and h^0 satisfies the interpolation conditions: $h^0_{|V^*} = 0$, and grad $h^0_{|V^*} = 0$. The interpolant h^* associated with h^0 is identically zero which follows from its expression in (3.1). Since $D^{2,1}h^0 = 0$ on I^2, it follows by (4.1) that h^0 is identically zero. Hence, $h^1 = h^2$.

Peano kernel representation of the interpolation errors $f - h^*$ and $f - h_*$. We shall need the following basic univariate error representation for Hermite–Lagrange interpolation (see [7] or [8] for details): *For $f \in C^m(I)$ let $P \in \pi_{m-1}$ be the algebraic polynomial to f of degree $m - 1$ satisfying $P(x_k) = f(x_k)$, $k = 1, 2, \ldots, m$ at some points $-1 \leq x_1 \leq \cdots \leq x_m \leq 1$ (at the coinciding knots we mean concurrence of the consecutive derivatives, i.e. Hermite interpolation). Then for $x \in I$*

$$f(x) - P(x) = \prod_{k=1}^{m}(x - x_k) \int_{-1}^{1} B_{m-1}(x_1, \ldots, x_m, x; t)f^{(m)}(t)dt, \qquad (4.3)$$

where $B_{m-1}(\,\cdot\,; t)$ is a B–spline of degree $m - 1$ with $m + 1$ knots $x_1, \ldots, x_m, x$ normalized with $\int_{-1}^{1} B_{m-1}(\cdot, t)dt = 1/m!$.

Theorem 6. *Let $f \in C^{2,1}(I^2)$. The following error representations hold true:*

(a) $\quad f(x, y) - h^*(x, y) = \displaystyle\int_{-1}^{1}\int_{-1}^{1} K^*(x, y; \xi, \eta)f^{(2,1)}(\xi, \eta)d\xi d\eta,$ $\qquad (4.4)$

$\quad$ *where* $\quad K^*(x, y; \xi, \eta) := (y - 2|x| + 1)\left[x^2 - ((\eta + 1)/2)^2\right] \times$

$\qquad \times B_0(y, 2|x| - 1; \eta)B_1\left(-(\eta + 1)/2, (\eta + 1)/2, x; \xi\right)$

$\quad$ *is the corresponding Peano kernel.*

(b) $\quad f(x,y) - h_*(x,y) = \displaystyle\int_{-1}^{1}\int_{-1}^{1} K_*(x,y;\xi,\eta)f^{(2,1)}(\xi,\eta)d\xi d\eta\ ,$ $\qquad$ (4.5)

where $K_*(x,y;\xi,\eta) := (y + 2|x| - 1)\left[x^2 - ((1-\eta)/2)^2\right]$

$\qquad\qquad \times B_0(y, -2|x| + 1;\eta)B_1\left(-(1-\eta)/2, (1-\eta)/2, x;\xi\right)$

is the corresponding Peano kernel.

Proof. Proof of (a). Denote $g(x,y) := f(x,y) - h^*(x,y)$. Using the fact that $g_{|V^*} = 0$ and grad $g_{|V^*} = 0$ we obtain

$$g(x,y) = (y - 2x + 1)\int_{-1}^{1} B_0(y, 2x - 1;\eta)g^{(0,1)}(x,\eta)d\eta.\ \text{for } (x,y) \notin V^*,\ x > 0\,.$$

On the other hand

$$g^{(0,1)}(x,\eta)$$
$$= \left(x^2 - ((\eta+1)/2)^2\right)\int_{-1}^{1} B_1\left(-(\eta+1)/2, (\eta+1)/2, x;\xi\right) f^{(2,1)}(\xi,\eta)d\xi.$$

Hence, we have

$$g(x,y) \;=\; \int_{-1}^{1}\int_{-1}^{1}(y - 2x + 1)\left(x^2 - ((\eta+1)/2)^2\right)$$
$$\times B_0(y, 2x - 1;\eta)B_1\left(-(\eta+1)/2, (\eta+1)/2, x;\xi\right) f^{(2,1)}(\xi,\eta)\,d\xi\,d\eta.$$

Combining this formula with the analogous formula for $x < 0$ we get (a) for $(x,y) \in I^2$.

The proof of (b) is analogous to that of (a).

Remark 5. Since $h^*_{f(u,-v)}(x,-y) = h_{*f(u,v)}(x,y)$, by changing y to $-y$ in

$$f(x,-y) - h^*_{f(u,-v)}(x,y) = -\int_{-1}^{1}\int_{-1}^{1} K^*(x,y;\xi,-\eta)f^{(2,1)}(\xi,\eta)d\xi d\eta\ ,$$

we obtain

$$f(x,y) - h_{*f(u,v)}(x,y) = -\int_{-1}^{1}\int_{-1}^{1} K^*(x,-y;\xi,-\eta)f^{(2,1)}(\xi,\eta)d\xi d\eta\ ,$$

and from here $K_*(x,y;\xi,\eta) = -K^*(x,-y;\xi,-\eta)$.

5   Cubature Formulae Exact for $B^{2,1}\left(I^2\right)$

Observing that $xF^*(y)$ and $\int_0^x F^*(2|t|-1)\,dt$ are odd functions with respect to x and integrating the interpolant h^* from Theorem 3 on I^2, we obtain

$$
\int_{I^2} h^* = 2\int_{-1}^{1}\int_0^x f^{(1,0)}(t,2|t|-1)dtdx + \int_{-1}^{1}\int_{-1}^{y} f^{(0,1)}((t+1)/2,t)dtdy
$$
$$
+ \int_{-1}^{1}\int_{-1}^{y} f^{(0,1)}(-(t+1)/2,t)dtdy + 4f(0,-1)
$$
$$
= 2\int_{-1}^{1}\int_0^x \left[f^{(1,0)}(t,2|t|-1) + 2\operatorname{sign}(t)f^{(0,1)}(t,2|t|-1)\right]dtdx + 4f(0,-1)
$$
$$
= 2\int_{-1}^{1}\int_0^x \frac{d}{dt}f(t,2|t|-1)dtdx + 4f(0,-1)
$$
$$
= 2\int_{-1}^{1} f(x,2|x|-1)dx .
$$

Integrating (4.1) on I^2, we obtain

$$
\int_{I^2} f(x,y)dxdy = \int_{I^2} h^*(x,y)dxdy - \frac{1}{3}D^{2,1}f(\rho^*,\sigma^*),
$$

where $(\rho^*,\sigma^*) \in I^2$. Hence, we get the cubature formula

$$
\int_{I^2} f(x,y)dxdy = 2\int_{-1}^{1} f(x,2|x|-1)dx - \frac{1}{3}D^{2,1}f(\rho^*,\sigma^*) \qquad (5.1)
$$

with $(\rho^*,\sigma^*) \in I^2$. The cubature (5.1) is exact for any $h \in B^{2,1}(I^2)$. Analogously, integrating the error representation (4.2) on I^2 we get the cubature formula

$$
\int_{I^2} f(x,y)dxdy = 2\int_{-1}^{1} f(x,-2|x|+1)dx + \frac{1}{3}D^{2,1}f(\rho_*,\sigma_*), \qquad (5.2)
$$

where $(\rho_*,\sigma_*) \in I^2$. The cubature (5.2) is exact for $h \in B^{2,1}(I^2)$. Denote

$$
CF^*(f) := 2\int_{-1}^{1} f(x,2|x|-1)dx \quad \text{and} \quad CF_*(f) := 2\int_{-1}^{1} f(x,-2|x|+1)dx.
$$

Peano kernel representations for the cubature formulae (5.1) *and* (5.2). Here we determine the Peano kernels K^* and K_* such that the representations

$$\int_{I^2} f - 2\int_I f(x, 2|x| - 1)dx = \int_{I^2} K^*(\xi, \eta)f^{(2,1)}(\xi, \eta)d\xi d\eta \qquad (5.3)$$

and

$$\int_{I^2} f - 2\int_I f(x, -2|x| + 1)dx = \int_{I^2} K_*(\xi, \eta)f^{(2,1)}(\xi, \eta)d\xi d\eta. \qquad (5.4)$$

hold. Since

$$K^*(\xi, \eta) = \int_{I^2} K^*(x, y; \xi, \eta)dx dy \quad \text{and} \quad K_*(\xi, \eta) = \int_{I^2} K_*(x, y; \xi, \eta)dx dy$$

it follows that (see Remark 5)

$$K_*(\xi, \eta) = -K^*(\xi, -\eta).$$

Hence, it is sufficient to obtain a representation for $K^*(\xi, \eta)$.

Let $\overline{h}$ denote the $B^{2,1}$–interpolant to $f \in C^{2,1}(I^2)$ such that $\overline{h}$ coincides with f on the sets $\{(-1, y) \mid y \in [-1, 1]\}$, $\{(1, y) \mid y \in [-1, 1]\}$ and $\{(x, -1) \mid x \in [-1, 1]\}$, see [6] for details. We have

$$f(x, y) - \overline{h}(x, y) = (x^2 - 1)(y + 1)$$
$$\times \int_{I^2} B_1(-1, x, 1; \xi)B_0(-1, y; \eta)f^{(2,1)}(\xi, \eta)d\xi d\eta$$

By the explicit forms of B_0 and B_1 we get

$$f(x, y) - \overline{h}(x, y) = \frac{1}{2}\int_{I^2} \overline{K}(x, y; \xi, \eta)f^{(2,1)}(\xi, \eta)d\xi d\eta ,$$

where we have (using truncated power functions, see [7] or [8] for details)

$$\overline{K}(x, y; \xi, \eta) := [-(1 + x)(1 - \xi) + 2(x - \xi)_+](y - \eta)^0_+.$$

Therefore

$$\int_{I^2} f - \int_{I^2} \overline{h} = \frac{1}{2}\int_{I^2} \left(\int_{I^2} \overline{K}(x, y; \xi, \eta)dx dy\right) f^{(2,1)}(\xi, \eta)d\xi d\eta.$$

Further we calculate

$$\int_{I^2} \overline{K}(x, y; \xi, \eta)dx dy = -(1 - \eta)(1 - \xi^2)$$

and

$$
CF^*(\overline{K})(\xi,\eta) = -4(1-\xi)\int_{\frac{1+\eta}{2}}^{1} dx + 4\int_{\max(\xi,\frac{1+\eta}{2})}^{1}(x-\xi)dx
$$

$$
+4\int_{0}^{1}(2x-1-\eta)_{+}^{0}(-x-\xi)_{+}dx.
$$

We consider three cases concerning the relation of ξ and $\frac{1+\eta}{2}$.

(a) $\frac{1+\eta}{2} \leq \xi \leq 1$. In this case we have (with equality only for $\eta = -1$ or $\xi = 1$)

$$
K^*(\xi,\eta) = (1/2)\left[\left(\int_{I^2}\overline{K}\right)(\xi,\eta) - CF^*(\overline{K})(\xi,\eta)\right] = -(1/2)(1-\xi)^2(1+\eta) \leq 0.
$$

(b) $-\frac{1+\eta}{2} \leq \xi \leq \frac{1+\eta}{2}$. In this case (with equality for $\eta = 1$ or $(\xi,\eta) = (0,-1)$)

$$
\begin{aligned}
K^*(\xi,\eta) &= (1/2)\left[\left(\int_{I^2}\overline{K}\right)(\xi,\eta) - CF^*(\overline{K})(\xi,\eta)\right] \\
&= -(1/2)\left[(1-\xi)^2(1+\eta) - \frac{1}{2}(1+\eta-2\xi)^2\right] \leq 0.
\end{aligned}
$$

(c) $-1 \leq \xi \leq -\frac{1+\eta}{2}$. In this case we obtain (with equality only for $\eta = -1$ and $\xi = -1$)

$$
K^*(\xi,\eta) = (1/2)\left[\left(\int_{I^2}\overline{K}\right)(\xi,\eta) - CF^*(\overline{K})(\xi,\eta)\right] = -(1/2)(1+\eta)(1+\xi)^2 \leq 0.
$$

Finally, summing up we get the following explicit expression for the non-positive kernel $K^*(\xi,\eta)$

$$
K^*(\xi,\eta) = \begin{cases} -\frac{1}{2}(1+\eta)(1-|\xi|)^2\,, & (1+\eta)/2 \leq |\xi| \leq 1, \\[2mm] \frac{1}{2}(1-\eta)\left(\xi^2 - (1+\eta)/2\right)\,, & |\xi| \leq (1+\eta)/2. \end{cases}
$$

6   Proofs of Theorems 1 and 2

Proof of Theorem 1. By the assumption $D^{2,1}f \geq 0$ we get from Theorem 5 (a) that $h^* \geq f$ on I^2.

Let $h \in B^{2,1}(I^2)$ and $h \geq f$ on I^2. We have

$$\int_{I^2} (h - f) = \int_{I^2} (h - h^*) + \int_{I^2} (h^* - f).$$

By the cubature (5.1), which is exact for functions from the linear space $B^{2,1}(I^2)$, we obtain

$$\int_{I^2} (h - h^*) = 2 \int_{-1}^{1} (h - h^*)(x, 2|x| - 1)\, dx \;=\; 2 \int_{-1}^{1} (h - f)(x, 2|x| - 1)\, dx \geq 0,$$

since $h \geq f$ in I^2 and $h^*_{|V^*} = f_{|V^*}$. From here

$$\int_{I^2} (h - f) \geq \int_{I^2} (h^* - f) \tag{6.1}$$

for each $h \in B^{2,1}(I^2)$ satisfying $h - f \geq 0$. This means that h^* is a best one–sided L^1–approximant to f from above with respect to $B^{2,1}(I^2)$.

Suppose that there exists a point $(x_0, y_0) \in V^*$ such that

$$h(x_0, y_0) > f(x_0, y_0) = h^*(x_0, y_0).$$

Then $\int_{I^2}(h - h^*) = 2 \int_{-1}^{1} (h - h^*)(x, 2|x| - 1)\, dx > 0$, and from here

$$\int_{I^2} (h - f) > \int_{I^2} (h^* - f),$$

thus h is not a best one–sided approximant to f from above. It follows that equality in (6.1) may occur only if

$$h_{|V^*} = f_{|V^*} = h^*_{|V^*}. \tag{6.2}$$

Next we use the following consequence of Taylor's formula : *Let v and w be two continuously differentiable univariate functions in some neighborhood $U(a)$ of*

$a \in \mathbb{R}$, let $v(a) = w(a)$ and $v(x) \geq w(x)$ in $U(a)$. Then $v'(a) = w'(a)$. Thus by $h \geq f$ on I^2 and (6.2) it follows that

$$\operatorname{grad} h_{|V^*} = \operatorname{grad} f_{|V^*} = \operatorname{grad} h^*_{|V^*}.$$

By Theorem 3 it follows that $h = h^*$. Hence, $h^* \in B^{2,1}(I^2)$ is the unique best one-sided approximation to f from above.

Remark 6. Besides of this elementary proof, we can prove Theorem 1 also using a general result due to Armitage and Gardiner [2, Proposition 1]. This will be done in the

Proof of Theorem 2.

From Theorem 5 (b) we see that $h^* \leq f$. For $D^{2,1} f > 0$, from Theorem 5 (b) we have $Z(f - h^*) := \{(x, y) \in I^2 \mid (f - h^*)(x, y) = 0\} = \Lambda_*$. Hence the best one–sided approximation assertion of Theorem 2 is an immediate consequence of Proposition 1 by Armitage and Gardiner [2] and the cubature formula (5.2) which is exact for $f \in B^{2,1}(I^2)$. In the case when $D^{2,1} f \geq 0$, an easy perturbation argument gives the assertion. Uniqueness follows as in the proof of Theorem 1.

Remark 7. Similar results can be obtained in the case of best one–sided L^1–approximation by $B^{1,2}$–blending functions with the following canonical grids for $B^{1,2}$–blending interpolation: $<^* := \{(x, y) \in I^2 : x = 2|y| - 1\}$ when we approximate from above and $>_* := \{(x, y) \in I^2 : x = -2|y| + 1\}$ for approximation from below.

Acknowledgments. This research was partially supported by the Sofia University Grant 328/2000–2001 and by the Grant I/70523 of the Volkswagen Foundation.

References

[1]   N. I. Achieser: *Theory of Approximation*, Frederic Ungar, New York, 1956.

[2]   D. H. Armitage, S. J. Gardiner: *Best one–sided L^1–approximation by harmonic and subharmonic functions*, In: *Advances in Multivariate Approximation, Mathematical Research* Vol. **107**, 43–56, Wiley–VCH, Berlin 1999.

[3]   D. H. Armitage, S. J. Gardiner, W. Haußmann, L. Rogge: *Characterization of best harmonic and superharmonic L¹–approximants,* J. reine angew. Math. **478** (1996), 1–15.

[4]   D. H. Armitage, S. J. Gardiner, W. Haußmann, L. Rogge: *Best one–sided L¹–approximation by harmonic functions,* Manuscripta math. **96** (1998), 181–194.

[5]   R. Bojanić, R. A. DeVore: *On polynomials of best one–sided approximation,* L'Enseignement Math. **12** (1966), 139–164.

[6]   B. D. Bojanov, D. P. Dryanov, W. Haußmann, G. P. Nikolov: *Best one–sided L¹–approximation by blending functions.* In: *Advances in Multivariate Approximation,* Mathematical Research Vol. **107**, 85–106, Wiley–VCH, Berlin 1999.

[7]   B. D. Bojanov, H. A. Hakopian, A. A. Sahakian: *Spline Functions and Multivariate Interpolations,* Mathematics and its Applications Vol. **248**, Kluwer, Dordrecht 1993.

[8]   C. de Boor: *A Practical Guide to Splines,* Applied Mathematical Sciences Vol. **27**, Springer, New York–Berlin 1978.

[9]   R. A. DeVore: *One–sided approximation of functions,* J. Approx. Theory **1** (1968), 11–25.

[10]   R. A. DeVore, G.G. Lorentz: *Constructive Approximation,* Springer, Berlin 1993.

[11]   D. Dryanov, W. Haußmann, P. Petrov: *Best one–sided L¹–approximation of bivariate functions by sums of univariate ones,* Arch. Math. (Basel) **75** (2000), 125–131.

[12]   W. Haußmann, K. Zeller: *Blending interpolation and best L¹–approximation.* Arch. Math. (Basel) **40** (1983), 545–552.

[13]   W. Haußmann, K. Zeller: *Canonical point sets in multivariate constructive function theory,* In: *Israel Math. Conf. Proc.* Vol. **11**, 91–104, Bar–Ilan Univ., Ramat Gan, 1997.

[14]   A. Pinkus: *On L¹-Approximation.* Cambridge University Press
            1989.

Addresses:

Dimiter Dryanov
Département des Mathématiques et de Statistique
Université de Montréal, C. P. 6128
Montréal, Québec H3C 3J7
Canada

Werner Haußmann
Mathematisches Institut
Gerhard–Mercator–Universität
D–47048 Duisburg
Germany

Petar Petrov
Department of Mathematics
University of Sofia
BG–1164 Sofia
Bulgaria

International Series of Numerical Mathematics
Vol. 137, ©2001 Birkhäuser Verlag Basel/Switzerland

Open Problem: Existence of Hermite Interpolatory Subdivision Schemes with Arbitrary Large Smoothnesses

Nira Dyn

In the case of "Lagrange" interpolatory subdivision schemes, it is known that the family of Deslauriers–Dubuc schemes [2] furnishes a sequence of schemes with increasing supports and increasing smoothnesses [1]. To be more specific, a member of this family is an interpolatory subdivision scheme L_N of the form $f^{k+1} = L_N f^k$, given by the rules

$$f^{k+1}_{2\alpha} = f^k_\alpha, \quad \alpha \in \mathbb{Z}, \quad k \geq 0 \tag{1}$$

$$f^{k+1}_{2\alpha+1} = P_{2N,\alpha}(\tfrac{1}{2}), \quad \alpha \in \mathbb{Z}, \quad k \geq 0, \tag{2}$$

where $P_{2N,\alpha}$ is a polynomial of degree $2N - 1$, satisfying the Lagrange interpolation conditions:

$$P_{2N,\alpha}(j) = f^k_{\alpha+j}, \quad j = -N+1, -N+2, \ldots, -1, 0, 1, \ldots, N \tag{3}$$

It follows from (2) and (3) that

$$f^{k+1}_{2\alpha+1} = \sum_{j=-N+1}^{N} a_j f^k_{\alpha+j}$$

and thus the support of the scheme L_N grows linearly with N. It is known [1] that the smoothness of the functions generated by L_N also grows linearly with N, but with a much smaller rate of growth.

A "natural" family $\{H_N\}$ of Hermite interpolatory subdivision schemes, which is an extension of $\{L_n\}$, is defined by $F^{k+1} = H_N F^k$ with $F^k_\alpha = (f^k_\alpha, g^k_\alpha)^T$, given by the rules

$$F^{k+1}_{2\alpha} = F^k_\alpha, \quad \alpha \in \mathbb{Z}, \quad k \geq 0 \tag{4}$$

$$F^{k+1}_{2\alpha+1} = \begin{pmatrix} P_{4N,\alpha}(\tfrac{1}{2}) \\ 2^{-k} P'_{4N,\alpha}(\tfrac{1}{2}) \end{pmatrix}, \tag{5}$$

where $P_{4N,\alpha}$ is a polynomial of degree $4N - 1$, satisfying the Hermite interpolation conditions

$$P_{4N,\alpha}(j) = f_{\alpha+j}^k, \quad P'_{4N,\alpha}(j) = 2^k g_{\alpha+j}^k, \quad j = -N+1, \ldots, 0, 1, \ldots, N . \quad (6)$$

It follows from (5) and (6) that

$$F_{2\alpha+1}^{k+1} = \sum_{j=-N+1}^{N} A_j F_{\alpha+j}^k , \quad (7)$$

where $\{A_j\}_{j=-N+1}^{N}$ are 2×2 matrices.

The function $H_N^\infty F^0$, generated by H_N from the initial data $F^0 = \{F_\alpha^0 : \alpha \in \mathbb{Z}\}$, is a function which is at least in C^1 and which satisfies

$$H_N^\infty F^0(2^{-k}\alpha) = f_\alpha^k, \quad (H_N^\infty F^0)'(2^{-k}\alpha) = 2^k g_\alpha^k, \quad \alpha \in \mathbb{Z}, \quad k \in \mathbb{Z}_+ . \quad (8)$$

Although the support of H_N, determined by (5), (6) and (7), grows linearly with N, it is not known, except from numerical evidence for small values of N [3], whether the smoothness of the functions generated by H_N grows with N unboundedly.

The open problem is to prove the

Conjecture. *For any $m \in \mathbb{Z}_+$, there exists $N \in \mathbb{Z}_+$ such that the functions generated by H_N are in C^m.*

This result is needed in [4], but a weaker result suffices: *For any $m \in \mathbb{Z}_+$, there exists a Hermite interpolatory subdivision scheme generating C^m functions.* For more information about Hermite interpolatory subdivision schemes, see [5] and [6].

References

[1]   I. Daubechies: *Ten Lectures on Wavelets,* Soc. Ind. Appl. Math., Philadelphia 1992.

[2]   G. Deslauriers, S. Dubuc: *Interpolation dyadique,* in: Fractals, Dimension Non Entieres et Applications, G. Cherbit (ed.), Masson, Paris 1987, pp. 44–55.

[3]   D. Donoho, N. Dyn, D. Levin, T. Yu: *Smooth multiwavelet duals of Alpert bases by moment–interpolating refinement,* Appl. Comput. Harm. Anal. **9** (2000), 166–203.

[4]   N. Dyn: *A construction of bi–orthogonal functions to B–splines with multiple knots,* Appl. Comput. Harm. Anal. **8** (2000), 24–31.

[5]   N. Dyn, D. Levin: *Analysis of Hermite–type subdivision schemes,* in: Approximation Theory VIII–Wavelets and Multilevel Approximation, C. K. Chui, L. L. Schumaker (eds.), World Sci. Publ., Singapore 1995, pp. 117–124.

[6]   N. Dyn, D. Levin: *Analysis of Hermite–interpolatory subdivision schemes,* in: Spline Functions and the Theory of Wavelets, S. Dubuc (ed.), AMS series CRM Proceedings and Lecture Notes **18**, Providence, R. I., 1999, pp. 105–113.

Address:

Nira Dyn
School of Mathematical Sciences
Tel Aviv University
Tel Aviv, 69978
Israel

International Series of Numerical Mathematics
Vol. 137, ©2001 Birkhäuser Verlag Basel/Switzerland

Asymptotic Formulas in Cardinal Interpolation and Orthogonal Projection

Karol Dziedziul

Abstract

We consider asymptotic formulas for the error in cardinal interpolation and orthogonal projection. Here we prove that the limit of these formulas are connected with Bernoulli–Stöckler splines. We introduce the concept of bootstrap approximations which follows from a weak saturation theorem for orthogonal projection.

1 Introduction

Let us consider the broken line interpolation I_h of the function $f(x) = \sin^9 x$ on the interval $[0,6]$ with break points at $hk \in [0,6]$ and $k \in \mathbb{N}$. The rate of convergence is h^2. Figures 1 (resp. 2) show both

$$\frac{I_h f(x) - f(x)}{h^2},$$

where $h = 1/8$ (resp. $h = 1/16$) and the function $f''/12$.

This is a visualization of the phenomenon, which is proved in [2], that if $h \to 0$ then

$$\frac{I_h f(x) - f(x)}{h^2} \to \frac{f''}{12},$$

where the convergence is weak in $L^2(\mathbb{R})$. We will prove in Section 5 that such phenomena can be observed in general.

A natural question is: What is the behaviour of

$$\left\| \frac{I_h f(x) - f(x)}{h^2} \right\|_p \to ?$$

For $p = 2$ this was considered in [10], also in [2] with the weak saturation theorem described above. For $1 \leq p < \infty$ results will appear in [4].

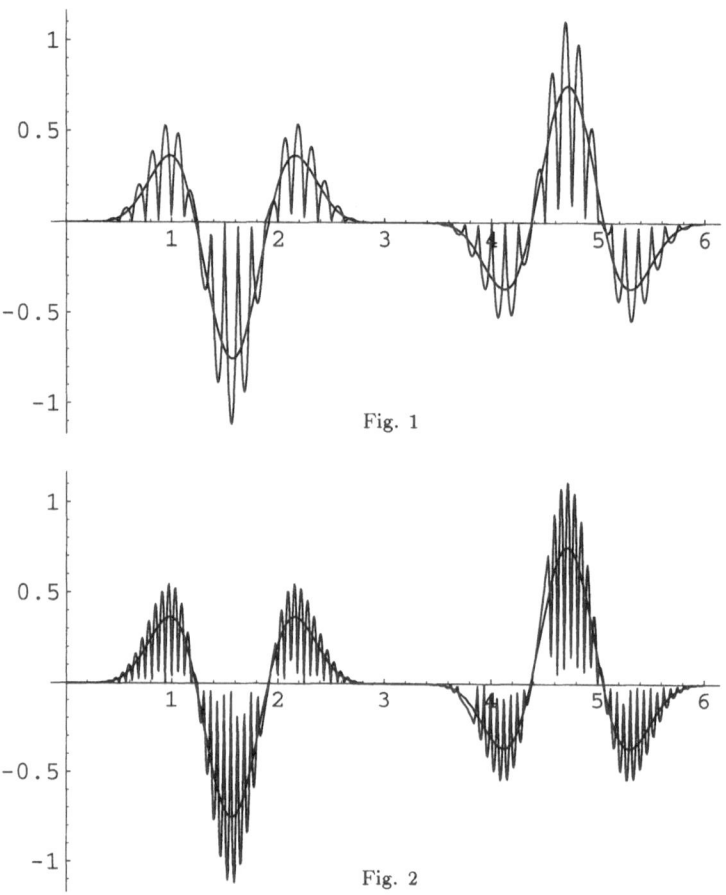

Fig. 1

Fig. 2

For the orthogonal projection on shift invariant spaces a similar result will be published in [5]. In Section 3 we will calculate the right hand side of the asymptotic formula in box spline shift invariant spaces using Bernoulli–Stöckler splines. The orthogonal projections have another interesting property proved in [3] which we call weak saturation theorem. For functions f smooth enough

$$\frac{P_h f - f}{h^{\rho v + 1}} \to 0.$$

where P_h is defined by (2.6) and where the convergence is weak in $L^2(\mathbb{R}^d)$. This leads us to *bootstrap approximation or smoothing decoding*, see Section 4. In fact there are operators Q_h such that for a sufficiently smooth function f we have

$$\|Q_h(P_h f) - f\|_p = O(h^{2\rho v + 2}). \tag{1.1}$$

2 Notation

We start by recalling some notations and basic facts.

Let $W_p^k(\mathbb{R}^d)$ denote the Sobolev spaces, $1 \le p < \infty$ with the norm

$$\|f\|_{k,p} = \sum_{|\beta| \le k} \|D^\beta f\|_p,$$

where

$$D^\beta f = \frac{\partial^{|\beta|} f}{\partial x_1^{\beta_1} \cdots \partial x_d^{\beta_d}}, \quad \beta = (\beta_1, \cdots, \beta_d), \quad |\beta| = \beta_1 + \cdots + \beta_d.$$

and

$$\|f\|_p = \left(\int_{\mathbb{R}^d} |f|^p \right)^{1/p}.$$

We use the standard convolution notation

$$(f * g)(x) = \int_{\mathbb{R}^d} f(y)g(x - y)\, dy.$$

Let $V = \{v_1, v_2, \cdots, v_n\}$ denote a family of not necessarily distinct vectors in $\mathbb{Z}^d \setminus \{0\}$, such that

$$\text{span}\{V\} = \mathbb{R}^d.$$

Such a set V is called admissible. The box spline corresponding to V will be denoted by $B_V(\cdot)$ and is defined by the condition that

$$\int_{\mathbb{R}^d} f(x) B_V(x)\, dx = \int_{[0,1]^n} f(Vu)\, du \tag{2.1}$$

holds for any continuous function f on $\mathbb{R}^d$.

The Fourier transform is given by

$$\hat{f}(\xi) = \int_{\mathbb{R}^d} f(t) e^{-2\pi i \xi \cdot t}\, dt.$$

Here and subsequently the dot denotes the scalar product in $\mathbb{R}^d$. From (2.1) a simple calculation leads to

$$\hat{B}_V(\xi) = \prod_{v \in V} \frac{1 - e^{-2\pi i \xi \cdot v}}{2\pi i \xi \cdot v}. \tag{2.2}$$

We denote by $|X|$ the cardinality of the set X. For an admissible set V define ρ_V as

$$\rho_V = \max\{r : \operatorname{span}\{V \setminus W\} = \mathbb{R}^d \text{ for all } W \subset V \text{ with } |W| = r\}.$$

This parameter determines the smoothness of a box spline:

$$B_V(\cdot) \in C^{\rho_V - 1}(\mathbb{R}^d) \setminus C^{\rho_V}(\mathbb{R}^d).$$

Moreover, if Π_k denotes the space of all polynomials in d variables of total degree at most k then

$$\Pi_{\rho_V} \subset \operatorname{span}\{B_V(\cdot - \alpha) : \alpha \in \mathbb{Z}^d\}.$$

Let us define

$$S_{L^2}(hV) = \overline{\operatorname{span}\{B_V(\cdot/h - \alpha) : \alpha \in \mathbb{Z}^d\}},$$

where $h > 0$ and where the closure is taken in $L^2(\mathbb{R}^d)$ topology. The orthogonal projection from $L^2(\mathbb{R}^d)$ onto $S_{L^2}(hV)$ is denoted by P_h.

By $-V$ we denote the family of vectors

$$-V := \{-v : v \in V\}.$$

A family V is called unimodular if for all $W \subset V$ such that $|W| = d$ we have $|\det W| \leq 1$. Put

$$X = V \cup -V.$$

The following conditions are equivalent:

i) V *is unimodular.*

ii) *For all* $x \in \mathbb{R}^d$

$$w(x) = \sum_{\alpha \in \mathbb{Z}^d} B_X(\alpha) e^{2\pi i \alpha \cdot x} \neq 0. \tag{2.3}$$

Moreover in this case the sequence $\{b(\alpha)\}$ of the Fourier expansion of the function $\dfrac{1}{w(x)}$ is symmetric

$$b(\alpha) = b(-\alpha) \tag{2.4}$$

and decays exponentially, i.e. there are constants $C > 0$ and $0 < q < 1$ such that for all $\alpha \in \mathbb{Z}^d$

$$|b(\alpha)| \le Cq^{|\alpha|}.$$

The function

$$B_V^*(x) = \sum_{\alpha \in \mathbb{Z}^d} b(\alpha)B_V(x - \alpha) \qquad (2.5)$$

is biorthogonal i.e.

$$\int_{\mathbb{R}^d} B_V^*(t)B_V(t - \alpha)\, dt = \begin{cases} 0 & \text{for } \alpha \ne 0, \\ 1 & \text{for } \alpha = 0. \end{cases}$$

Denoting by $(\cdot, \cdot)$ the inner product in $L^2(\mathbb{R}^d)$, the orthogonal projection P_h onto $S_{L^2}(hV)$ can be written by

$$P_h = \sigma_h \circ P \circ \sigma_{\frac{1}{h}}, \qquad (2.6)$$

where

$$\sigma_h f(x) = f\left(\frac{x}{h}\right)$$

and

$$Pf(x) = \sum_{\alpha \in \mathbb{Z}^d} (f, B_V^*(\cdot - \alpha))B_V(x - \alpha)$$

is the orthogonal projection onto $S_{L^2}(V)$. In the considered case we can introduce the fundamental function

$$\Phi(x) = \sum_{\alpha \in \mathbb{Z}^d} b(\alpha)B_X(x - \alpha), \qquad (2.7)$$

which can also be expressed as

$$\Phi(x) = \int_{\mathbb{R}^d} B_V^*(s)B_V(x + s)ds. \qquad (2.8)$$

By definition we have

$$\Phi(\alpha) = \begin{cases} 0 & \alpha \ne 0, \\ 1 & \alpha = 0, \end{cases} \qquad (2.9)$$

for $\alpha \in \mathbb{Z}^d$. Moreover, suppose that there are constants $C > 0$ and $0 < q < 1$ such that

$$|\Phi(x)| < Cq^{\|x\|},$$

where $x \in \mathbb{R}^d$ and $\|x\| = \sqrt{x \cdot x}$. Thus we can define the cardinal interpolation

$$If(x) = \sum_{\alpha \in \mathbb{Z}^d} f(\alpha)\Phi(x - \alpha).$$

We have

$$I_h = \sigma_h \circ I \circ \sigma_{\frac{1}{h}}, \quad h > 0 \tag{2.10}$$

or equivalently

$$I_h f(x) = \sum_{\alpha \in \mathbb{Z}^d} f(h\alpha)\Phi(\frac{x}{h} - \alpha).$$

Note that

$$\rho_X = 2\rho_V + 1.$$

For $|\beta| \le \rho_V + 1$ we define the periodic functions

$$L_\beta(x) = \sum_{\alpha \in \mathbb{Z}^d} B_V^*(x - \alpha) \int_{\mathbb{R}^d} B_V(s - \alpha)(x - s)^\beta \, ds,$$

and for $|\beta| \le \rho_X + 1$

$$K_\beta(x) = \sum_{\alpha \in \mathbb{Z}^d} (x - \alpha)^\beta \Phi(x - \alpha).$$

3 Asymptotic Formulas

Let us recall the results from [4], [5]. They are concerned with the case $1 \le p < \infty$. It emerges that using the limit theorem we can prove them also for $p = \infty$. Best constants are connected with Bernoulli numbers.

Theorem 3.1. *Let $1 \le p < \infty$. Let V be unimodular and $f \in W_p^{\rho_V+1}(\mathbb{R}^d)$. Then*

$$\lim_{h \to 0^+} \left\| \frac{f - P_h f}{h^{\rho_V+1}} \right\|_p^p = \tag{3.1}$$

$$= \int_{\mathbb{R}^d} \Big(\int_{[0,1]^d} \Big| \sum_{|\beta|=\rho_V+1} \frac{1}{\beta!} D^\beta f(t) L_\beta(x) \Big|^p \, dx \Big) dt.$$

Theorem 3.2. *Let $1 \leq p < \infty$. Let V be unimodular and $f \in C_0^r(\mathbb{R}^d)$. Then*

$$\lim_{h \to 0+} \left\| \frac{f - I_h f}{h^{\rho_X + 1}} \right\|_p^p =$$

$$\int_{\mathbb{R}^d} \left(\int_{[0,1]^d} \left| \sum_{|\beta| = \rho_X + 1} \frac{1}{\beta!} D^\beta f(t) K_\beta(x) \right|^p dx \right) dt. \tag{3.2}$$

Put

$$[]^\beta(x) = x^\beta$$

and

$$\gamma \leq \beta \quad \text{if and only if} \quad \gamma_j \leq \beta_j, j = 1, \ldots, d.$$

Lemma 3.3. *Let $0 < |\beta| \leq \rho_X + 1$. Then*

$$K_\beta(x) = (-1)^{|\beta|}(I([]^\beta)(x) - x^\beta). \tag{3.3}$$

Let $0 < |\beta| \leq \rho_V + 1$. Then

$$L_\beta(x) = (-1)^{|\beta|}(P([]^\beta)(x) - x^\beta) \tag{3.4}$$

Proof. Since for $|\gamma| < \rho_X + 1$ the interpolation operator I restricted to Π_{ρ_X} is the identity, it follows that

$$\sum_{\alpha \in \mathbb{Z}^d} \alpha^\gamma \Phi(x - \alpha) = I([]^\gamma)(x) = x^\gamma.$$

Then, by definition,

$$K_\beta(x) = \sum_{0 \leq \gamma \leq \beta} \binom{\beta}{\gamma} (-1)^{|\gamma|} x^{\beta - \gamma} \sum_{\alpha \in \mathbb{Z}^d} \alpha^\gamma \Phi(x - \alpha) =$$

$$= \sum_{0 \leq \gamma < \beta} \binom{\beta}{\gamma} (-1)^{|\gamma|} x^{\beta - \gamma} x^\gamma + (-1)^{|\beta|} I([]^\beta)(x) =$$

$$= (-1)^{|\beta| + 1} x^\beta + (-1)^{|\beta|} I([]^\beta)(x).$$

This finishes the proof of (3.3). We turn to the proof of (3.4). The orthogonal projection P from $L^2(\mathbb{R}^d)$ to $S_{L^2}(V)$ is well defined also for polynomials. Moreover, P is the identity on Π_{ρ_V}. Thus for $|\gamma| < \rho_V + 1$ we have

$$\sum_{\alpha \in \mathbb{Z}^d} B_V^*(x - \alpha) \int_{\mathbb{R}^d} B_V(s - \alpha) s^\gamma \, ds = P([]^\gamma)(x) = x^\gamma.$$

By definition

$$L_\beta(x) = \sum_{\alpha \in \mathbb{Z}^d} B_V^*(x - \alpha) \sum_{0 \le \gamma \le \beta} (-1)^{|\gamma|} \binom{\beta}{\gamma} x^{\beta-\gamma} \int_{\mathbb{R}^d} B_V(s - \alpha) s^\gamma \, ds =$$

$$= \sum_{0 \le \gamma < \beta} (-1)^{|\gamma|} \binom{\beta}{\gamma} x^{\beta-\gamma} P([]^\gamma)(x) + (-1)^{|\beta|} P([]^\beta)(x) =$$

$$= (-1)^{|\beta|+1} x^\beta + (-1)^{|\beta|} P([]^\beta)(x).$$

$\square$

Note that (3.4) gives in the univariate case that L_β is a Bernoulli spline, see [18], [19].

Lemma 3.4. *Let $0 < |\beta| \le \rho_X + 1$. Then*

$$K_\beta(x) = \left(-\frac{1}{2\pi i}\right)^{|\beta|} D^\beta \widehat{\Phi}(0) + \left(-\frac{1}{2\pi i}\right)^{|\beta|} \sum_{\alpha \in \mathbb{Z}^d, \alpha \ne 0} D^\beta \widehat{B_X}(\alpha) \cdot e^{2\pi i \alpha \cdot x} \quad (3.5)$$

and

$$L_\beta(x) = \left(-\frac{1}{2\pi i}\right)^{|\beta|} \sum_{\alpha \in \mathbb{Z}^d, \alpha \ne 0} D^\beta \widehat{B_V}(\alpha) \cdot e^{2\pi i \alpha \cdot x}. \quad (3.6)$$

Proof. Applying the Poisson formula for the function $[]^\beta \Phi$ we get

$$\sum_{\alpha \in \mathbb{Z}^d} (x - \alpha)^\beta \Phi(x - \alpha) = \left(-\frac{1}{2\pi i}\right)^{|\beta|} \sum_{\alpha \in \mathbb{Z}^d} D^\beta \widehat{\Phi}(\alpha) \cdot e^{2\pi i \alpha \cdot x}, \quad (3.7)$$

where by definition of the fundamental function

$$\widehat{\Phi}(x) = \widehat{B_X}(x) Q(x), \quad (3.8)$$

and where

$$Q(x) = \sum_{\alpha \in \mathbb{Z}^d} b(\alpha) \cdot e^{-2\pi i \alpha \cdot x} = \sum_{\alpha \in \mathbb{Z}^d} b(\alpha) \cdot e^{2\pi i \alpha \cdot x}.$$

The last equation follows since the sequence $b(\alpha)$ is symmetric.

Strang–Fix conditions give (this can be seen by straightforward calculation)

$$D^\beta B_X(\alpha) = 0 \quad (3.9)$$

for $\alpha \neq 0$ and $|\beta| < \rho_X + 1$. Thus since

$$Q(0) = \sum_{\alpha \in \mathbb{Z}^d} b(\alpha) = \frac{1}{\sum_{\alpha \in \mathbb{Z}^d} B_X(\alpha)} = 1,$$

we get for $\alpha \neq 0$

$$D^\beta \widehat{\Phi}(\alpha) = \sum_{0 \leq \gamma \leq \beta} \binom{\beta}{\gamma} D^\gamma \widehat{B_X}(\alpha) D^{\beta-\gamma} Q(\alpha) = \qquad (3.10)$$

$$= \begin{cases} 0 & \text{if } |\beta| < \rho_X + 1, \\ D^\beta \widehat{B_X}(\alpha) & \text{if } |\beta| = \rho_X + 1. \end{cases}$$

(3.10) and (3.7) give (3.5). Let us turn to the proof of (3.6). Let us define

$$g_\beta(x) := B_V^*(x) \int_{\mathbb{R}^d} B_V(u)(x-u)^\beta du =$$

$$\sum_{0 \leq \gamma \leq \beta} \binom{\beta}{\gamma} (-1)^{|\beta-\gamma|} x^\gamma B_V^*(x) \int_{\mathbb{R}^d} B_V(u) u^{\beta-\gamma} du.$$

Note that

$$L_\beta(x) = \sum_{\alpha \in \mathbb{Z}^d} g_\beta(x-\alpha).$$

and that

$$\widehat{g_\beta}(t) = \sum_{0 \leq \gamma \leq \beta} \binom{\beta}{\gamma} (-1)^{|\beta-\gamma|} \left(\frac{1}{-2\pi i}\right)^{|\gamma|} D^\gamma \widehat{B_V^*}(t) \int_{\mathbb{R}^d} B_V(u) u^{\beta-\gamma} du.$$

By definition (2.5)

$$\widehat{g_\beta}(t) = \sum_{0 \leq \gamma \leq \beta} \binom{\beta}{\gamma} (-1)^{|\beta-\gamma|} \left(\frac{1}{-2\pi i}\right)^{|\gamma|} D^\gamma \left(\widehat{B_V} Q\right)(t) \int_{\mathbb{R}^d} B_V(u) u^{\beta-\gamma} du.$$

Now, applying the Poisson formula for the function g_β and using the formula (3.9) and (3.10) – note that B_V^* plays the role of Φ and B_X is replaced by B_V – we get

$$L_\beta(x) = \sum_{0 \leq \gamma \leq \beta} \binom{\beta}{\gamma} (-1)^{|\beta-\gamma|} \left(\frac{1}{-2\pi i}\right)^{|\gamma|} \int_{\mathbb{R}^d} B_V(u) u^{\beta-\gamma} du \times$$

$$\sum_{\alpha \in \mathbb{Z}^d} D^\gamma \left(\widehat{B_V} Q\right)(\alpha) \cdot e^{2\pi i \alpha \cdot x} =$$

$$= \sum_{0 \leq \gamma \leq \beta} \binom{\beta}{\gamma} (-1)^{|\beta-\gamma|} \int_{\mathbb{R}^d} B_V(u) u^{\beta-\gamma} du \left(\frac{1}{-2\pi i}\right)^{|\gamma|} D^\gamma \left(\widehat{B_V Q}\right)(0) +$$

$$\left(\frac{1}{-2\pi i}\right)^\beta \sum_{\alpha \in \mathbb{Z}^d, \alpha \neq 0} D^\beta \widehat{B_V}(\alpha) \cdot e^{2\pi i \alpha \cdot x}.$$

To finish the proof it is sufficient to show that the constant

$$C := \sum_{0 \leq \gamma \leq \beta} \binom{\beta}{\gamma} (-1)^{|\beta-\gamma|} \int_{\mathbb{R}^d} B_V(u) u^{\beta-\gamma} du \left(\frac{1}{-2\pi i}\right)^{|\gamma|} D^\gamma \left(\widehat{B_V Q}\right)(0)$$

is zero. We have

$$\int_{\mathbb{R}^d} B_V(u) u^{\beta-\gamma} du = \left(\frac{1}{-2\pi i}\right)^{|\beta-\gamma|} D^{\beta-\gamma} \widehat{B_V}(0).$$

This leads to

$$C = \sum_{0 \leq \gamma \leq \beta} \binom{\beta}{\gamma} (-1)^{|\beta-\gamma|} \left(\frac{1}{-2\pi i}\right)^{|\beta-\gamma|} \left(\frac{1}{-2\pi i}\right)^{|\gamma|} D^{\beta-\gamma} \widehat{B_V}(0) D^\gamma \left(\widehat{B_V Q}\right)(0)$$

$$= \left(\frac{1}{-2\pi i}\right)^{|\beta|} \sum_{0 \leq \gamma \leq \beta} \binom{\beta}{\gamma} D^{\beta-\gamma} \widehat{B_{-V}}(0) D^\gamma \left(\widehat{B_V Q}\right)(0) =$$

$$= \left(\frac{1}{-2\pi i}\right)^{|\beta|} D^\beta \left(\widehat{B_{-V}} \widehat{B_V Q}\right)(0) =$$

$$= \left(\frac{1}{-2\pi i}\right)^{|\beta|} D^\beta \widehat{\Phi}(0).$$

The following Lemma 3.5 shows that the constant C vanishes. $\qquad\square$

Lemma 3.5. *For $0 < |\beta| < \rho_X + 1$ we have*

$$D^\beta \widehat{\Phi}(0) = 0. \tag{3.11}$$

For $|\beta| = \rho_X + 1$ we get

$$K_\beta(x) = \sum_{\alpha \in \mathbb{Z}^d} (x - \alpha)^\beta B_X(x - \alpha) - \sum_{\alpha \in \mathbb{Z}^d} \alpha^\beta B_X(\alpha). \tag{3.12}$$

Proof. For $0 < |\beta| < \rho_X + 1$ equation (3.3) implies $K_\beta \equiv 0$, thus by (3.5) and (3.9) we get (3.11). To prove (3.12) let us note that from (3.3) and (2.9) we have

$$K_\beta(0) = \sum_{\alpha \in \mathbb{Z}^d} \alpha^\beta \Phi(-\alpha) = 0. \tag{3.13}$$

But if we apply the Poisson formula for $[]^\beta B_X$ and use the formula (3.5) we get

$$K_\beta(x) = \left(-\frac{1}{2\pi i}\right)^{|\beta|} (D^\beta \widehat{\Phi}(0) - D^\beta \widehat{B_X}(0)) + \sum_{\alpha \in \mathbb{Z}^d} (x - \alpha)^\beta B_X(x - \alpha).$$

This and (3.13) give (3.12).

$\square$

To simplify the considerations let $U \subset \mathbb{Z}^2 \setminus \{0\}$ such that for the given family V of vectors

$$v_1 = \begin{pmatrix} 1 \\ 0 \end{pmatrix}, \quad v_2 = \begin{pmatrix} 0 \\ 1 \end{pmatrix}, \quad v_3 = \begin{pmatrix} 1 \\ 1 \end{pmatrix},$$

with multiplicities $\mu = (\mu_1, \mu_2, \mu_3)$ we have

$$\alpha \cdot v_j \neq 0 \quad \text{for all} \quad j = 1, ..., k, \text{ and all } \alpha \in U.$$

Let us recall the Bernoulli–Stöckler splines [16]:

$$B_V^\mu(U)(x) = \sum_{\alpha \in U} \prod_{j=1}^3 (i\alpha \cdot v_j)^{-\mu_j} \cdot e^{2\pi i \alpha \cdot x}.$$

These considerations lead to

Theorem 3.6. *Let V be a family of the vectors*

$$v_1 = \begin{pmatrix} 1 \\ 0 \end{pmatrix}, \quad v_2 = \begin{pmatrix} 0 \\ 1 \end{pmatrix}, \quad v_3 = \begin{pmatrix} 1 \\ 1 \end{pmatrix},$$

with multiplicities μ_1, μ_2, μ_3. The functions L_β are finite linear combinations of Bernoulli–Stöckler splines.

Proof. We will calculate the functions L_β. For $x = (x_1, x_2) \in \mathbb{R}^2$ it follows

$$\widehat{B_V}(x) = \prod_{j=1}^3 [f(v_j \cdot x)]^{\mu_j},$$

where
$$f(t) = \frac{1 - e^{-2\pi it}}{2\pi it}.$$

Let
$$r := \rho_V + 1.$$

Then for all $i \neq j$
$$\mu_i + \mu_j \geq r. \tag{3.14}$$

Moreover there are $i \neq j$ such that in (3.14) we have equality. Moreover let us notice that if $\alpha \cdot v_j \neq 0$ then

$$D^{\delta_j}[f]^{\mu_j}(\alpha \cdot v_j) = \begin{cases} \mu_j! \left(\frac{1}{\alpha \cdot v_j}\right)^{\mu_j} & \delta_j = \mu_j, \\ 0 & \delta_j < \mu_j. \end{cases} \tag{3.15}$$

Let us introduce the BS splines. For a triple $(\delta_1, \delta_2, \delta_3) \in \mathbb{Z}^3$ such that $\delta_1 + \delta_2 + \delta_3 = r$, define

$$BS(x) = BS(\delta_1, \delta_2, \delta_3)(x) = \sum_{\alpha \neq 0} \prod_{j=1}^{3} D^{\delta_j}[f(\alpha \cdot v_j)]^{\mu_j} \cdot e^{2\pi i \alpha \cdot x}. \tag{3.16}$$

If $BS \neq 0$ then $\delta_{i_1} = 0$, $\mu_{i_2} = \delta_{i_2}$ and $\mu_{i_3} = \delta_{i_3}$. Then from (3.15) we have that

$$BS(x) = \sum_{\alpha \neq 0, v_{i_1} \cdot \alpha = 0} \mu_{i_2}! \mu_{i_3}! \left(\frac{1}{v_{i_2} \cdot \alpha}\right)^{\mu_{i_2}} \left(\frac{1}{v_{i_3} \cdot \alpha}\right)^{\mu_{i_3}} \cdot e^{2\pi i \alpha \cdot x}.$$

From (3.6), (3.16) and from the definition of Bernoulli–Stöckler splines it follows that the functions L_β consist of linear combinations of Bernoulli–Stöckler splines.

$\square$

Example 3.7. In fact if $\mu_1 = \mu_2 = \mu_3 = 2$ then $r = 4$ and

$$L_{(4,0)} = \frac{6}{8\pi^4} BS(2, 0, 2),$$

$$L_{(3,1)} = \frac{3}{8\pi^4} BS(2, 0, 2),$$

$$L_{(2,2)} = \frac{1}{8\pi^4} (BS(2, 0, 2) + BS(2, 2, 0) + BS(0, 2, 2)),$$

$$L_{(1,3)} = \frac{3}{8\pi^4} BS(0, 2, 2),$$

$$L_{(0,4)} = \frac{6}{8\pi^4} BS(0, 2, 2).$$

4 Bootstrap Approximation

We can double the rate of convergence of the orthogonal projection (2.6) using for instance the convolution operator with the fundamental function Φ (2.7). In fact let

$$Q_h f = f * \Phi_h,$$

where $\Phi_h(x) = \dfrac{1}{h^d} \Phi\left(\dfrac{x}{h}\right)$ and

$$Qf = f * \Phi.$$

Then

Theorem 4.1. *If* $f \in W_p^{2\rho v + 2}$*, then*

$$\|Q_h(P_h f) - f\|_p = O(h^{2\rho v + 2}). \tag{4.1}$$

Proof. Note that

$$Q_h = \sigma_h \circ Q \circ \sigma_{1/h}$$

and

$$Q_h P_h = \sigma_h \circ Q \circ P \circ \sigma_{1/h}$$

Thus to prove this theorem it is sufficient to show that for all polynomials $p \in \Pi_{2\rho v + 1}$

$$Q(P(p)) = p.$$

Note that

$$Q(P(p))(x) = \sum_{\alpha \in \mathbb{Z}^d} p * (\check{B}_V^*)(\alpha)\Phi * B_V(x - \alpha),$$

where

$$(\check{B}_V^*)(x) = B_V^*(-x).$$

But $p * (\check{B}_V^*)$ is a polynomial of degree less than or equal to $2\rho v + 1$. Moreover from (2.7)

$$B_X(x) = \sum_{\alpha \in \mathbb{Z}^d} B_X(\alpha)\Phi(x - \alpha),$$

hence

$$\Pi_{2\rho v + 2} \subset \text{span}\{\Phi * B_V(\cdot - \alpha) : \alpha \in \mathbb{Z}^d\}.$$

Thus the de Boor–Ron equality implies that we can replace discrete convolution by ordinary convolution (see [14]):

$$Q(P(p))(x) = p * (\check{B}_V^*) * \Phi * B_V(x).$$

Now (2.8) and the de Boor–Ron equality for Φ give

$$Q(P(p)) = p * \Phi * \Phi = p * \Phi = p,$$

since the interpolation restricted to $\Pi_{\rho X}$ is the identity.   □

Theorem 4.2. *If $f \in W_2^{2\rho v+2}$ then*

$$\left\|\frac{Q_h(P_h f) - f}{h^{2\rho v+2}} - \sum_{|\beta|=2\rho v+2} \left(\frac{1}{2\pi i}\right)^{2\rho v+2} \frac{A_\beta}{\beta!} D^\beta f(x)\right\|_2 = o(1), \quad h \to 0, \quad (4.2)$$

where

$$A_\beta = D^\beta \widehat{B_X}(0) - (2\pi i)^{2\rho v+2} \sum_{\alpha \in \mathbb{Z}^d} \alpha^\beta B_X(\alpha).$$

Proof. It is known that the convolution operator defined by a function Φ with the property

$$D^\beta \widehat{\Phi}(0) = 0, \quad 0 < |\beta| < \rho_X + 1,$$

has a convergence rate h^{ρ_X+1}. But $\rho_X + 1 = 2\rho_V + 2$. Thus it is sufficient to consider

$$\|Q_h(P_h f) - Q_h f\|_2.$$

From Plancherel's formula we get

$$\|Q_h(P_h f) - Q_h f\|_2^2 = \int_{\mathbb{R}^d} \left|\widehat{\Phi}(hx)[\widehat{P_h f}(x) - \widehat{f}(x)]\right|^2 dx. \quad (4.3)$$

Let us take a sufficient smooth function $\widehat{f}$ (we apply Poisson's formula) with compact support $C = \operatorname{supp}\widehat{f}$. We split the integral into two integrals over C and $\mathbb{R}^d \setminus C$. Since $\widehat{\Phi}(0) = 1$, the calculations given in the proofs of Lemmas 2.3 and 2.4 in [3] give the asymptotic formula for sufficiently small h

$$\int_C \left|\widehat{\Phi}(hx)[\widehat{P_h f}(x) - \widehat{f}(x)]\right|^2 dx = \quad (4.4)$$

$$= \int_C \left|\widehat{\Phi}(hx)\widehat{f}(x)[\widehat{B_V}(hx)\widehat{B_{-V}}(hx)/w(-hx) - 1]\right|^2 dx \sim h^{2(\rho_X+1)}.$$

The same proof as of Lemma 2.4 in [3] gives

$$\frac{1}{h^{\rho_X+1}}[\widehat{B_V}(hx)\widehat{B_{-V}}(hx)/w(-hx) - 1] \to \quad (4.5)$$

$$\sum_{|\beta|=\rho_X+1} \frac{x^\beta}{\beta!} \left[D^\beta \widehat{B_X}(0) - (2\pi i)^{\rho_X+1} \sum_{\alpha \in \mathbb{Z}^d} \alpha^\beta B_X(\alpha) \right]$$

uniformly for $x \in C$.

On the other hand for sufficiently small h formula (2.20) in [3] gives (using $\widehat{\Phi}(x) = \widehat{B_X}(x)w(x)$)

$$\int_{\mathbb{R}^d \setminus C} \left| \widehat{\Phi}(hx) \widehat{P_h f}(x) \right|^2 dx = \tag{4.6}$$

$$= \sum_{\gamma \neq 0} \int_C \left| \widehat{B_V}(hx+\gamma) \widehat{B_X}(hx+\gamma)w(hx) \widehat{B_{-V}}(hx) \widehat{f}(x)/w(-hx) \right|^2 dx.$$

The calculations (2.21)–(2.24) in [3] give the needed asymptotic formula

$$\sum_{\gamma \neq 0} \left| \widehat{B_V}(hx+\gamma) \widehat{B_X}(hx+\gamma) \right|^2 \sim h^{2(\rho_V + \rho_X + 2)}. \tag{4.7}$$

Equations (4.3)–(4.7) yield (4.2). By a standard argument we can obtain the theorem for all f. $\qquad\square$

5 An Asymptotic Formula for the Error of Linear Operators

Only for simplicity we consider the linear operator

$$Qf(x) = \sum_{\alpha \in \mathbb{Z}^d} \langle \lambda, f(\cdot + \alpha) \rangle B_V(x - \alpha), \tag{5.1}$$

where λ is a distribution with compact support and V is a family of vectors from $\mathbb{Z}^d$ such that $\rho_V = 1$. From the general Strang–Fix theorem we conclude that the maximum rate of convergence is h^2 for

$$Q_h = \sigma_h \circ Q \circ \sigma_{\frac{1}{h}}. \tag{5.2}$$

Let

$$\langle \check{\lambda}, f \rangle := \langle \lambda, f(-\cdot) \rangle.$$

To achieve the rate h^2 we have to assume that the function

$$H(x) := B_V * \check{\lambda}(x) = \langle \lambda, B_V(x + \cdot) \rangle \tag{5.3}$$

satisfies

$$\text{(i)} \quad \int_{\mathbb{R}^d} H(x)dx = 1,$$

and

$$\text{(ii)} \quad \int_{\mathbb{R}^d} x_j H(x)dx = 0$$

for all $j = 1, ..., d$, $x = (x_1, ..., x_d)$, and further that there are i, j such that

$$\text{(iii)} \quad \int_{R^d} x_j x_i H(x)dx \neq 0.$$

It is known (see [2, Theorem 2.3]) that the assumptions imposed on H imply that for smooth functions f we have

$$\frac{H_h * f - f}{h^2} \to D_1 f = \frac{1}{4\pi^2} \sum_{|\beta|=2} \frac{D^\beta \widehat{H}(0)}{\beta} D^\beta f \tag{5.4}$$

in the L^2 norm, where $H_h(x) = \frac{1}{h^d} H(\frac{x}{h})$.

Lemma 5.1. *Under above assumptions (i) – (iii) on H for a function $f \in W_2^2(\mathbb{R}^d)$ we get*

$$\frac{Q_h * f - f}{h^2} \to D_1 f, \tag{5.5}$$

where the convergence is weak in L^2, and

$$\left\| \frac{Q_h * f - f}{h^2} \right\|_2^2 \to \|D_2 f\|_2^2 + \|D_1 f\|_2^2, \tag{5.6}$$

where

$$\|D_2 f\|_2^2 = \left(\frac{1}{4\pi^2}\right)^2 \sum_{U \in \Lambda} \|D_U f\|_2^2 \sum_{\substack{\alpha \perp (V \setminus U) \\ \alpha \neq 0}} \prod_{v \in U} \frac{1}{(\alpha \cdot v)^2} \,,$$

and where $\Lambda = \{U \subset V : |U| = 2, \text{span}\{V \setminus U\} \neq \mathbb{R}^d\}$, $D_U = \prod_{u \in U} D_u$ where D_u is the derivative in direction u.

Proof. From the calculation given in [3, formula (2.3) with $\rho_V = 1$] for a sufficiently smooth (to apply Poisson's formula) function $\widehat{f}$ with compact support $C = \text{supp} \widehat{f}$

$$\widehat{Q_h f}(x) = \widehat{B_V}(hx) \sum_{\alpha \in \mathbb{Z}^d} \widehat{\lambda}(hx + \alpha)\widehat{f}(x + \alpha/h).$$

For small h and a function $\hat{g} \in L^2$ with compact support we get

$$\int \frac{Q_h * f - f}{h^2} g = \int \frac{\widehat{B_V}(hx)\widehat{\lambda}(hx)\hat{f}(x) - \hat{f}(x)}{h^2} \overline{\hat{g}(x)}dx =$$

$$= \int \frac{H_h * f - f}{h^2} g.$$

Thus (5.4) implies (5.5). To prove (5.6) we proceed as in proof of (2.3) in [3]. Since

$$\int \left|\frac{Q_h * f - f}{h^2}\right|^2 =$$

$$\int_C \left|\frac{\widehat{B_V}(hx)\widehat{\lambda}(hx)\hat{f}(x) - \hat{f}(x)}{h^2}\right|^2 dx + \int_{\mathbb{R}^d \backslash C} \left|\frac{\widehat{Q_h f}}{h^2}\right|^2,$$

it is sufficient to examine

$$\int_{\mathbb{R}^d \backslash C} \left|\frac{\widehat{Q_h f}}{h^2}\right|^2 = \sum_{\gamma \neq 0} \int_C \left|\frac{B_V(hu + \gamma)\widehat{\lambda}(hu)\hat{f}(u)}{h^2}\right|^2 du.$$

The assumption (i) gives that $\widehat{\lambda}(0) = 1$. Thus if $h \to 0$ then $\widehat{\lambda}(hu) \to 1$. The asymptotic formula for

$$\sum_{\gamma \neq 0} \int_C \left|\frac{B_V(hu + \gamma)\hat{f}(u)}{h^2}\right|^2 du$$

is given in [3]. □

References

[1]   M. Beśka, K. Dziedziul: *Multiresolution approximation and Hardy spaces*, J. Approx. Theory **88** (1997), 154–167.

[2]   M. Beśka, K. Dziedziul: *Saturation theorems for interpolation and the Bernstein–Schnabl operator*, Math. Comp. **70** (2001), 705–717.

[3]   M. Beśka, K. Dziedziul: *The saturation theorem for orthogonal projection*, in: Advances in Multivariate Approximation, W. Haußmann, K. Jetter and M. Reimer (eds.), Math. Research Vol. **107**, Wiley–VCH, Berlin 1999, pp. 73–83.

[4] M. Beśka, K. Dziedziul: *Asymptotic formula for the error in cardinal interpolation,* to appear in Numer. Math.

[5] M. Beśka, K. Dziedziul: *Asymptotic formula for the error in orthogonal projection,* to appear in Math. Nachr.

[6] C. de Boor, R. A. DeVore, A. Ron: *Approximation orders of FSI spaces in $L_2(\mathbb{R}^d)$,* Constr. Approx. **14** (1998), 631–652.

[7] C. de Boor, R. A. DeVore, A. Ron: *The structure of finitely generated shift–invariant spaces in $L_2(R^d)$,* J. Funct. Anal. **119** (1994), 37–78.

[8] C. de Boor, R. A. DeVore, A. Ron: *Approximation from shift-invariant subspaces of $L_2(\mathbb{R}^d)$,* Trans. Amer. Math. Soc. **341** (1994), 787–806.

[9] C. de Boor, K. Höllig, S. Riemenschneider: *Box Splines,* Springer 1993.

[10] I. Daubechies, M. Unser: *On the approximation power of convolution–based least squares versus interpolation,* IEEE Trans. Sign. Proc. **45**, (1997), 1697–1711.

[11] K. Dziedziul: *The saturation theorem for quasiprojections,* Stud. Sci. Math. Hungarica **35** (1999), 99–111.

[12] K. Dziedziul: *Box Splines,* polish version, Wydawnictwo Politechniki Gdańskiej, 1997.

[13] R. Q. Jia, C. A. Micchelli: *Using the refinement equations for the construction of pre–wavelets PP: power of two,* in: Curves and Surfaces, P. J. Laurent, A. Le Méhauté, L. L. Schumaker (eds.), Academic Press, New York 1991, pp. 209–246.

[14] A. Ron, N. Sivakumar: *The approximation order of box spline spaces,* Proc. Amer. Math. Soc. **117** (1993), 473–482.

[15] E. M. Stein: *Singular Integrals and Differentiability Properties of Functions,* Princeton Univ. Press, Princeton, N. J., 1970.

[16] J. Stöckler: *Interpolation mit mehrdimensionalen Bernoulli-Splines und periodischen Box-Splines,* Ph. D. dissertation, Duisburg 1988.

[17]   G. Strang, G. Fix: *A Fourier analysis of the finite element variational method,* C.I.M.E. II, Ciclo Erice (1971), in: Constructive Aspects of Functional Analysis, G. Geymonat (ed.), Cremonese, Rome 1973, pp. 793–840.

[18]   W. Sweldens, R. Piessens: *Quadrature formulae and asymptotic error expansions for wavelet approximations of smooth functions,* SIAM J. Numer. Anal. **31** (1994), 1240–1264.

[19]   W. Sweldens, R. Piessens: *Asymptotic error expansion of wavelet approximations of smooth functions II,* Numer. Math. **68** (1994), 377–401.

Address:

Karol Dziedziul
Faculty of Applied Mathematics
Technical University
ul. Narutowicza 12/12
80–952 Gdańsk
Poland

International Series of Numerical Mathematics
Vol. 137, ©2001 Birkhäuser Verlag Basel/Switzerland

Note on d–Extremal Configurations for the Sphere in $\mathbb{R}^{d+1}$

Mario Götz and Edward B. Saff[†]

Abstract

It is shown that d–extremal configurations of N points on the unit sphere in $\mathbb{R}^{d+1}$, i.e., points minimizing energy with respect to the Riesz kernel $|x - y|^{-d}$, are asymptotically equidistributed as $N \to \infty$.

Consider the unit sphere $S^d := \{x \in \mathbb{R}^{d+1} : |x| = 1\}$ in $\mathbb{R}^{d+1}$, $d \geq 2$, and denote by σ the surface measures on S^d, normalized to have total mass 1. For given $s > 0$, the discrete s–energy of a system of N points $\omega_N = \{x_1, \ldots, x_N\} \subset S^d$ is given by

$$E_d(s, \omega_N) := \sum_{1 \leq i < j \leq N} \frac{1}{|x_i - x_j|^s} .$$

The paper [2] contains asymptotics of the *minimal s–energy*

$$\mathcal{E}_d(s, N) := \inf_{\omega_N} E_d(s, \omega_N) ,$$

where the infimum is taken over all N–point sets $\omega_N \subset S^d$ (see also [3], [4], [5]). Any set $\omega_N^* = \{x_1^{(N)}, \ldots, x_N^{(N)}\} \subset S^d$, for which this infimum is attained is called an *s–extremal configuration*. It is well–known that under the assumption $s < d$, each sequence of such s–extremal configurations is asymptotically equidistributed in the sense that the normalized discrete measures $\mu_{\omega_N^*}$ associating mass $1/N$ with each point $x_i^{(N)}$ converge to σ in the weak–star topology:

$$\lim_{N \to \infty} \frac{1}{N} \sum_{i=1}^{n} f(x_i^{(N)}) - \int f \, d\sigma \qquad (f \in C(S^d)) . \tag{1.1}$$

[†]The research of this author was supported, in part, by the U.S. National Science Foundation research grant DMS 9801677.

The key tool for a proof of this relation is a comparison of the normalized energy $\frac{2}{N(N-1)} \mathcal{E}_d(s, N)$ with the s–energy

$$\iint \frac{1}{|x - y|^s} \, d\sigma(x) \, d\sigma(y) \tag{1.2}$$

of the measure σ. However, for $s \geq d$ the integral in (1.2) becomes infinite, and the conventional methods fail. In this note we will show that the weak–star convergence (1.1) does also hold in the case $s = d$:

Theorem. *Each sequence* $(\omega_N^*)_{N \geq 2}$ *of d–extremal configurations on S^d is equidistributed in the sense of* (1.1).

The proof of this result is based on the following asymptotic behavior of the d–energy.

Theorem A (see [2, Theorem 3]). *The minimal d–energy satisfies*

$$\lim_{N \to \infty} \frac{1}{N^2 \log N} \mathcal{E}_d(d, N) = \frac{1}{2d} \gamma_d \,,$$

where

$$\gamma_d := \frac{\Gamma\left(\frac{d+1}{2}\right)}{\Gamma(\frac{d}{2}) \cdot \Gamma(\frac{1}{2})} \,.$$

Proof of the Theorem. For $t > 0$ and $z \in S^d$, denote by $C(z, t)$ the intersection of the closed ball of radius t centered at z with the sphere S^d. Such sets are called spherical caps. They can also be viewed as the intersection of the sphere with some closed half–space in $\mathbb{R}^{d+1}$. We remark that (see [2, (3.7)])

$$\int_{S^d \setminus C(z,t)} \frac{d\sigma(y)}{|z - y|^d} = -\gamma_d \log t + \mathcal{O}(1) \qquad \text{as } t \to 0. \tag{1.3}$$

Let $\omega_N^* = \{x_1^{(N)}, \ldots, x_N^{(N)}\} \subset S^d$ be a set of d–extremal points. Define

$$U_i(x) := \sum_{\substack{j=1 \\ j \neq i}}^{N} \frac{1}{|x - x_j^{(N)}|^d} \qquad (x \in S^d).$$

Then

$$\mathcal{E}_d(d, N) = E_d(d, \omega_N^*) = \frac{1}{2} \sum_{i=1}^{N} U_i(x_i^{(N)}), \tag{1.4}$$

and since ω_N^* is a d–extremal configuration,

$$U_i(x_i^{(N)}) \leq U_i(x) \qquad (x \in S^d). \tag{1.5}$$

For the moment, fix $r > 0$ (sufficiently small), and set

$$D_i(r) := S^d \setminus C(x_i^{(N)}, \frac{r}{N^{1/d}}), \qquad D(r) := \bigcap_{i=1}^{N} D_i(r).$$

Assume, contrary to the assertion of the Theorem, that the measures $\mu_{\omega_N^*}$ do not converge to σ in the weak–star topology. Then, by Helly's selection theorem, there exists some unit measure $\mu \neq \sigma$ on S^d, which is the weak–star limit of the measures $\mu_{\omega_N^*}$ along some subsequence $\Lambda \subset \mathbb{N}$. By the Cramér–Wold theorem (cf. [1]), which states that a probability measure on Euclidean space is uniquely determined by the values it gives to halfspaces, there is a closed spherical cap C such that $\mu(C) < \sigma(C)$. Thus, one can find an $\varepsilon > 0$ such that (after possibly passing to another subsequence) the cardinality of $\omega_N^* \cap C$ satisfies

$$\#(\omega_N^* \cap C) \leq N \left(\sigma(C) - \varepsilon \right) \qquad (N \in \Lambda). \tag{1.6}$$

Now, choose a second spherical cap $C' \subset C$ such that

$$\frac{\sigma(C) - \varepsilon}{\sigma(C')} < 1, \qquad \rho := \text{dist}(C', S^d \setminus C) > 0.$$

Taking into account (1.3) and (1.6), an integration over $C' \cap D(r)$ yields

$$\int_{C' \cap D(r)} U_i(x) \, d\sigma(x) \leq \sum_{\substack{j=1 \\ j \neq i}}^{N} \int_{C' \cap D_j(r)} |x - x_j^{(N)}|^{-d} d\sigma(x)$$

$$\leq \sum_{x_j^{(N)} \in C} \gamma_d \cdot [-\log(rN^{-1/d})] + \sum_{x_j^{(N)} \notin C} \frac{1}{\rho^d} + \mathcal{O}(N)$$

$$\leq (\sigma(C) - \varepsilon)N\gamma_d \cdot [-\log(rN^{-1/d})] + \mathcal{O}(N).$$

Consequently, by (1.4) and (1.5),

$$\mathcal{E}_d(d, N) \leq \frac{\sigma(C) - \varepsilon}{\sigma(C' \cap D(r))} \cdot \frac{\gamma_d}{2d} \cdot N^2 \cdot \log N + \log \frac{1}{r} \mathcal{O}(N^2). \tag{1.7}$$

On the other hand,

$$\sigma(C' \cap D(r)) \geq \sigma(C') - \sum_{i=1}^{N} \sigma\left(C\left(x_i^{(N)}, rN^{-1/d}\right)\right) \geq \sigma(C') - \frac{\gamma_d \, r^d}{d} \, ,$$

where we used the fact that $\sigma(C(x,t)) \leq \gamma_d t^d / d$ for all $x \in S^d$. Thus, we may choose $r = r(d)$ so small that

$$\frac{\sigma(C) - \varepsilon}{\sigma(C' \cap D(r))} < 1 \, .$$

Inserting this estimate into (1.7) yields a contradiction to the asymptotic behavior of the minimal d–energy according to Theorem A. $\qquad \square$

References

[1] H. Cramér, H. Wold: *Some theorems on distribution functions.* J. London Math. Soc. **11** (1936), 290–294.

[2] A. B. J. Kuijlaars, E. B. Saff: *Asymptotics for minimal discrete energy on the sphere.* Trans. Amer. Math. Soc. **350** (1998), 523–538.

[3] E. A. Rakhmanov, E. B. Saff, Y. M. Zhou: *Electrons on the sphere.* In: Computational Methods and Function Theory, R. M. Ali, St. Ruscheweyh, E. B. Saff (eds.), World Scientific, Singapore 1995, pp. 111–127.

[4] E. A. Rakhmanov, E. B. Saff, Y. M. Zhou: *Minimal discrete energy on the phere.* Math. Research Letters, **1** (1994), 647–662.

[5] E. B. Saff, A. B. J. Kuijlaars: *Distributing many points on a sphere.* Math. Intelligencer **19** (1997), 5–11.

Addresses:

Mario Götz
Kath. Univ. Eichstätt–Ingolstadt
Ostenstraße 26
85071 Eichstätt
Germany

Edward B. Saff
Inst. for Constructive Mathematics
Department of Mathematics
University of South Florida
4202 East Fowler Avenue
Tampa, FL 33620
U.S.A.

International Series of Numerical Mathematics
Vol. 137, ©2001 Birkhäuser Verlag Basel/Switzerland

Some Cubature Formulae Using Mixed Type Data

Vesselin Gushev and Geno Nikolov

Abstract

We study some cubature formulae for integrals on $I^2 = [-1,1]^2$ that use two types of information for the integrand: line integrals over either the boundary of I^2 or the coordinate axes, and evaluations at the points of a uniform grid. The error of these cubature formulae is analyzed, in particular the exact Peano constants are found for some classes of functions of low smoothness.

1 Introduction

Quadrature formulae are the most exploited tool for approximate calculation of definite integrals in the univariate case. Namely, the numerical value of

$$\ell[g] := \int\limits_{-1}^{1} g(x)\,dx$$

is approximated by a quadrature formula Q_n, i.e., a linear functional of the form

$$Q_n[g] := \sum_{\nu=1}^{n} a_\nu g(z_\nu), \quad -1 \le z_1 < \ldots < z_n \le 1.$$

The nodes $\{z_\nu\}_1^n$ and the weights $\{a_\nu\}_1^n$ are usually chosen to ensure that the remainder functional

$$R[Q_n; g] := \ell[g] - Q_n[g]$$

vanishes on the set of all algebraic polynomials up to certain degree s.

There is a complete theory devoted to the error analysis of quadrature formulae (see, e.g., [2], [4], [8], [10]). In particular, for each quadrature formula Q_n which calculates $\ell[g]$ to the exact value whenever g is polynomial of degree not exceeding $s - 1$, there exist unimprovable error estimates of the form

$$|R[Q_n; g]| \le c_{r,p}(Q_n)\|g^{(r)}\|_p, \quad r = 1, \ldots, s, \tag{1.1}$$

where $\|\cdot\|_p$ is the $L_p(I)$–norm. Hence, the error in approximating $\ell[g]$ by $Q_n[g]$ is estimated in terms of a certain norm of a single derivative of the integrand. The existing software packages for numerical integration in the univariate case exploit also properties as definiteness and monotonicity in quadrature approximations to obtain effective a posteriori error estimates and rules for termination of calculations (see, e.g., [5], [6], [7] and the literature cited therein).

It is natural to wonder whether these features of quadrature formulae can be reproduced in the bivariate case. Consider double integrals over a rectangular region (without loss of generality assumed to be $I^2 = [-1, 1] \times [-1, 1]$) :

$$L[f] := \iint\limits_{I^2} f(x, y)\, dx\, dy. \tag{1.2}$$

Usually, $L[f]$ is approximated by a cubature formula

$$L[f] \approx C[f] := \sum_{i=1}^{n_1} \sum_{j=1}^{n_2} a_{ij} f(t_i, \tau_j), \quad (t_i, \tau_j) \in I^2, \tag{1.3}$$

with the hope that the error

$$E[C; f] := L[f] - C[f] \tag{1.4}$$

is small in absolute value. But how small is it? Is it possible, as in the univariate case, to estimate $|E[C; f]|$ by a certain norm of, say,

$$f^{(r,s)}(x, y) := \frac{\partial^{r+s} f}{\partial x^r \partial y^s}(x, y) \ ?$$

The answer is in the negative. If an estimate of the form $|E[C; f]| \leq c\|f^{(r,s)}\|$ would be true, then evidently $C[f]$ would be exact for the class of functions

$$B^{r,s}(I^2) := \{f \in C^{r,s}(I^2) : f^{(r,s)}(x, y) = 0 \ \text{ for } (x, y) \in I^2\} . \tag{1.5}$$

However, $B^{r,s}(I^2)$ is an infinite dimensional linear space of functions (called blending functions), and no cubature formula of the form (1.3) exists, which is exact for all functions in $B^{r,s}(I^2)$.

In the present paper we propose a scheme for construction of cubature formulae for approximate calculation of $L[f]$ of a more general form than (1.3), which uses two types of information about the integrand: point evaluations and line integrals. Our method is based on interpolation by blending functions and briefly, is the following:

1. The integrand f is interpolated on a certain *blending grid* by a blending function Bf.

2. $L[f - Bf]$ is approximated by $C_n[f - Bf]$, where C_n is a cubature formula of customary type, i.e., involving point evaluations only. The latter can be, e.g., a product type cubature formula $C_n[h] = \sum_{i=1}^{n} \sum_{j=1}^{n} a_i a_j h(t_i, \tau_j)$, associated with a quadrature formula $Q_n[g] = \sum_{i=1}^{n} a_i g(t_i)$.

As a result, we obtain a cubature formula S_n of the form

$$S_n[f] := C_n[f] + L[Bf] - C_n[Bf], \tag{1.6}$$

which uses only few line integrals and, roughly, n^2 point evaluations. Since the exact calculation of the integrals is not always possible, it is assumed that only few line integrals are used for S_n, and a better approximation to $L[f]$ is realized by increasing n (hence, the number of point evaluations), while the number of line integrals (and the integrals themselves) remains fixed.

It turns out that cubature formulae (1.6) exhibit all the useful properties known for quadrature formulae and discussed above. The paper is organized as follows: In Section 2 we give some well–known facts about Peano kernel theory for quadratures, and present some quadrature formulae which use equispaced nodes and have asymptotically best Peano constants. In Section 3 we describe our construction of the cubature formulae (1.6) and prove relations between their Peano kernels and the Peano kernels of related quadrature formulae. In Sections 4 and 5 this theory is applied to the special cases when the blending grid consists of the boundary of I^2 and the portions of coordinate axes contained in I^2, respectively. In order to make possible the use of nested type data, all the cubature formulae involve evaluations at the points of a uniform grid. We construct best cubature formulae of quasi–Chebyshev type in the Sobolev classes $W_p^{1,1}(I^2)$, $p = 1, 2$, and definite (positive and negative) cubature formulae of order (2,2). For these cubature formulae, the exact Peano constants in the Sobolev classes $W_p^{r,r}(I^2)$, $r = 1, 2$, $p = 1, 2, \infty$, are found. For the cubature formulae (1.6) with C_n being the trapezium product rule we prove that, if the integrand f has a non-negative (or non-positive) derivative $f^{(2,2)}$ in I^2, then the sequence $\{S_{2^k n}[f]\}_{k=0}^{\infty}$ converges monotonically to $L[f]$ with respect to k. Section 6 contains some concluding remarks.

2   Peano Kernel Theory for Quadratures

For more details about Peano kernel theory we refer the reader to [2], [3], [4], [8], [10]. Here, we extract only those facts we shall need in the sequel.

For any $r \in \mathbb{N}_0$, the truncated power function of order r is defined by

$$x_+^r := (\max\{x,0\})^r \text{ if } r \in \mathbb{N}, \text{ and } x_+^0 = \begin{cases} 0 & \text{if} \quad x < 0, \\ 1/2 & \text{if} \quad x = 0, \\ 1 & \text{if} \quad x > 0. \end{cases}$$

Throughout the paper, π_m will designate the set of algebraic polynomials of degree not exceeding m. A classical result of Peano (see, e.g., [2]) says that if R is a linear functional defined on $C(I)$ which vanishes on π_{s-1}, and $g^{(s-1)}$ is absolutely continuous on I, then $R[f]$ can be represented as

$$R[g] = \int_{-1}^{1} K_s(t) g^{(s)}(t)\, dt, \quad K_s(t) = R\left[\frac{(\cdot - t)_+^{s-1}}{(s-1)!}\right]. \tag{2.1}$$

We shall use (2.1) mainly when R is the remainder term of a quadrature formula, i.e., $R = R[Q_n; \cdot] = \ell - Q_n$ with ℓ and Q_n as defined in Section 1. In this case K_s will be referred to as the s–th Peano kernel of Q_n, and will be denoted by $K_s(Q_n; t)$. Clearly, $K_s(Q_n; t)$ exists only if $R[Q_n; \cdot]$ vanishes on π_{s-1}. The quadrature formula Q_n is said to have algebraic degree of precision m (in short, $\text{ADP}(Q_n) = m$), if $R[Q_n; \cdot] \equiv 0$ on π_m but $R[Q_n; \cdot] \not\equiv 0$ on π_{m+1}.

For $t \in [-1, 1]$, explicit representations of $K_s(Q_n; t)$ are:

$$K_s(Q_n; t) = \frac{(1-t)^s}{s!} - \frac{1}{(s-1)!} \sum_{\nu=1}^{n} a_\nu (t_\nu - t)_+^{s-1},$$

and

$$K_s(Q_n; t) = (-1)^s \left[\frac{(t+1)^s}{s!} - \frac{1}{(s-1)!} \sum_{\nu=1}^{n} a_\nu (t - t_\nu)_+^{s-1}\right].$$

As usual, $\|\cdot\|_p$ means the $L_p(I)$–norm, i.e., $\|h\|_p := (\ell[|h|^p])^{1/p}$ if $1 \leq p < \infty$, and $\|h\|_\infty = \operatorname{ess\,sup}_{t \in I} |h(t)|$. The Sobolev classes $W_p^r(I)$ are defined by

$$W_p^r(I) := \{h \in C^{r-1}(I) : h^{(r-1)} \text{ absolutely continuous on } I, \|h^{(r)}\|_p < \infty\}.$$

If $g \in W_p^s(I)$ and $\text{ADP}(Q_n) \geq s - 1$, then Hölder's inequality yields the unimprovable error estimate

$$|R[Q_n; g]| \leq c_{s,p}(Q_n) \|g^{(s)}\|_p, \tag{2.2}$$

where $c_{s,p}(Q_n) = \|K_s(Q_n; \cdot)\|_q$, $p^{-1} + q^{-1} = 1$.

It should be mentioned that, in general, it is not easy to compute the exact value of the Peano constant $c_{s,p}(Q_n)$, even in the simplest cases $p = 1, 2$ and ∞. An exception for $p = \infty$ is the case when Q_n is a positive or negative definite quadrature formula of order s, i.e., when $K_s(Q_n; t) \geq 0$ or $K_s(Q_n; t) \leq 0$ on I. In this case, for $g \in C^s(I)$, the application of the mean value theorem yields $R[Q_n; g] = g^{(s)}(\xi)\ell[K_s(Q_n; \cdot)]$, in particular, $c_{s,\infty}(Q_n) = |R[Q_n; (\cdot)^s/s!]|$. The importance of definite quadrature formulae of order s stems from the fact that they provide one–sided approximation to $\ell[g]$ when $g^{(s)} \geq 0$ or $g^{(s)} \leq 0$ on I.

At the end of this section we present some asymptotically optimal quadrature formulae, i.e., formulae, whose Peano constants tend to the best possible as n goes to infinity.

Theorem 2.1. *Let $t_\nu = -1 + 2\nu/n$ for $\nu = 0, \ldots, n$. Then:*

(i) *The quadrature formula*

$$Q_n[g] = \frac{5}{6n}[g(t_0) + g(t_n)] + \frac{13}{6n}[g(t_1) + g(t_{n-1})] + \frac{2}{n}\sum_{\nu=2}^{n-2} g(t_\nu) \qquad (2.3)$$

is asymptotically optimal in $W_2^2[-1, 1]$, and

$$|R[Q_n; g]| \leq \left(\frac{2}{45n^4} + \frac{4}{27n^5}\right)^{1/2} \|g''\|_2.$$

(ii) *The quadrature formula*

$$Q_n[g] = \frac{13}{16n}[g(t_0) + g(t_n)] + \frac{35}{16n}[g(t_1) + g(t_{n-1})] + \frac{2}{n}\sum_{\nu=2}^{n-2} g(t_\nu) \qquad (2.4)$$

is asymptotically optimal in $W_\infty^2[-1, 1]$, and

$$|R[Q_n; g]| \leq \left(\frac{1}{4n^2} + \frac{533}{1536n^3}\right) \|g''\|_\infty.$$

(iii) *The quadrature formula*

$$Q_n^+[g] = \frac{15}{4n}[g(t_1) + g(t_{n-1})] + \frac{5}{4n}[g(t_2) + g(t_{n-2})] + \frac{2}{n}\sum_{\nu=3}^{n-3} g(t_\nu) \qquad (2.5)$$

$(n \geq 5)$ *is an asymptotically optimal positive definite quadrature formula of order* 2. *If* $g \in C^2[-1, 1]$, *then*

$$R[Q_n^+; g] = \left(\frac{1}{3n^2} + \frac{5}{n^3}\right) g''(\xi) \quad \text{for some} \quad \xi \in [-1, 1].$$

Proof. Between every two consecutive nodes x_ν and $x_{\nu+1}$ of a quadrature formula Q (exact for π_1) its second Peano kernel $K_2(Q; \cdot)$ is a polynomial of degree 2 with leading coefficient $1/2$, so the L_q deviation of $K_2(Q; \cdot)$ on $[x_\nu, x_{\nu+1}]$ is not smaller than the L_q–norm of the corresponding polynomial of degree 2 least deviating from zero. For $q = 2$ and $q = 1$ these are the Legendre and the Chebyshev polynomial of the second kind, respectively, given by

$$P_{2,\nu}(x) = \frac{1}{2}[(x - (x_\nu + x_{\nu+1})/2)^2 - (x_{\nu+1} - x_\nu)^2/12]$$

and

$$U_{2,\nu}(x) = \frac{1}{2}[(x - (x_\nu + x_{\nu+1})/2)^2 - (x_{\nu+1} - x_\nu)^2/16].$$

Therefore, we have

$$\|K_2(Q; \cdot)\|_{L_2[x_\nu, x_{\nu+1}]}^2 \geq \frac{(x_{\nu+1} - x_\nu)^5}{720}, \quad \|K_2(Q; \cdot)\|_{L_1[x_\nu, x_{\nu+1}]} \geq \frac{(x_{\nu+1} - x_\nu)^3}{32}.$$

If $x_1 < \ldots < x_m$ $(m \leq n)$ are the nodes of Q located in $(-1, 1)$, by summing up and taking into account $\sum_{\nu=1}^{N} \alpha_\nu^r \geq N^{-r+1}(\sum_{\nu=1}^{N} \alpha_\nu)^r$ with $N = m + 1$, $r = 3, 5$ and $\alpha_\nu = x_\nu - x_{\nu-1}$ $(x_0 := -1, x_{m+1} = 1)$, we get lower bounds showing the asymptotical optimality of the quadrature formulae in (i) and (ii):

$$\|K_2(Q; \cdot)\|_2^2 \geq \frac{2^5}{720(m + 1)^4} \geq \frac{2}{45(n + 1)^4},$$

$$\|K_2(Q; \cdot)\|_1 \geq \frac{2^3}{32(m + 1)^2} \geq \frac{1}{4(n + 1)^2}.$$

Equality would follow if $m = n$, $\{x_\nu\}_0^{n+1}$ are equispaced and $K_2(Q; \cdot)_{[x_\nu, x_{\nu+1}]}$ coincides with $P_{2,\nu}$ $(U_{2,\nu}$, respectively) for $\nu = 0, \ldots, n$. This goal cannot be achieved on the boundary intervals $[x_0, x_1]$ and $[x_n, x_{n+1}]$, as $K_2(Q; \pm 1) = 0$. However, as simple calculation shows, if the points $\{t_\nu\}_0^n$ are taken as nodes, then an appropriate choice of the weights corresponding to $g(t_0)$ and $g(t_1)$ yields $K_2(Q; \cdot)_{[t_1, t_2]} = P_{2,1}$ (or $= U_{2,1}$, respectively). Further, the weight $2/n$ for $\{g(t_\nu)\}_{\nu=2}^{n-2}$ guarantees $K_2(Q; \cdot)_{[t_\nu, t_{\nu+1}]} = P_{2,\nu}$ (or $= U_{2,\nu}$, respectively) for

$\nu = 2, \ldots, n - 2$. Finally, the weights of $g(t_{n-1})$ and $g(t_n)$ are chosen by symmetry. This procedure results in the quadrature formulae (2.3) and (2.4), respectively, and the corresponding Peano constants are easily calculated.

The same argument is used to derive (2.5): this time $K_2(Q; \cdot)|_{[x_\nu, x_{\nu+1}]}$ is compared with $U_{2,\nu}^+$, the least L_1-deviating on $[x_\nu, x_{\nu+1}]$ non-negative polynomial,

$$U_{2,\nu}^+(x) = \frac{1}{2}(x - \frac{x_\nu + x_{\nu+1}}{2})^2.$$

We thus get a lower bound showing that Q_n^+ is asymptotically optimal definite quadrature formula of order 2:

$$\|K_2(Q; \cdot)\|_1 \geq \sum_{\nu=0}^{m} \|U_{2,\nu}^+\|_{L_1[x_\nu, x_{\nu+1}]} = \frac{1}{24} \sum_{\nu=0}^{m} (x_{\nu+1} - x_\nu)^3 \geq \frac{1}{3(n+1)^2}.$$

We take $\{t_\nu\}_1^{n-1}$ as nodes and choose suitable weights for $g(t_1)$ and $g(t_2)$ (and equal corresponding weights of $g(t_{n-1})$ and $g(t_{n-2})$) to realize $K_2(Q_n^+; \cdot) \geq 0$ on $[t_0, t_2]$ and $K_2(Q_n^+; \cdot)|_{[t_2, t_3]} = U_{2,2}^+$. Then the weight $2/n$ for $\{g(t_\nu)\}_{\nu=3}^{n-3}$ ensures $K_2(Q_n^+; \cdot)|_{[t_\nu, t_{\nu+1}]} = U_{2,\nu}^+$ ($\nu = 3, \ldots, n - 3$). By symmetry, $K_2(Q_n^+; \cdot) \geq 0$ on $[t_{n-2}, t_n]$. The proof of (iii) is completed by calculation of $\|K_2(Q_n^+; \cdot)\|_1$. $\quad\square$

Remark 1. The quadrature formulae (2.3) and (2.5) as well as their asymptotical optimality are known in the literature. The first one is known as Durand's quadrature formula (see [2, p. 22]), while the second is due to Schmeißer [10]. The same arguments give the well-known fact that the trapezoid rule

$$Q_n^{tr}[g] = \frac{1}{n}[g(t_0) + g(t_n)] + \frac{2}{n} \sum_{\nu=1}^{n-1} g(t_\nu) \qquad (2.6)$$

is an optimal negative definite quadrature formula of order 2.

3 Cubature Formulae Using Mixed Type Data

We start with some facts about *blending functions* which can be found, e.g., in [1]. For $m_1, m_2 \in \mathbb{N}$, the space of blending functions $B^{m_1, m_2}(I^2)$ is defined by

$$B^{m_1, m_2}(I^2) := \{f \in C^{m_1, m_2}(I^2) \; : \; D^{m_1, m_2} f := \frac{\partial^{m_1 + m_2}}{\partial x^{m_1} \partial y^{m_2}} f = 0\}, \qquad (3.1)$$

where

$$C^{m,n}(I^2) := \{f : I^2 \to \mathbb{R} \; : \; D^{i,j} f \text{ continuous}, \; 0 \leq i \leq m, \; 0 \leq j \leq n\}.$$

Given two sets $X = \{x_1, x_2, \ldots, x_{m_1}\}$ and $Y = \{y_1, y_2, \ldots, y_{m_2}\}$ such that $-1 \leq x_1 < \ldots < x_{m_1} \leq 1$ and $-1 \leq y_1 \ldots < y_{m_2} \leq 1$), we define a blending grid $G = G(X, Y)$ by

$$G(X, Y) := \{(x, y) \in I^2 \ : \ \prod_{\mu=1}^{m_1}(x - x_\mu) \prod_{\nu=1}^{m_2}(y - y_\nu) = 0\}.$$

For any function f defined in I^2 there exists a unique (Lagrange) blending interpolant $Bf = B_G f \in B^{m_1, m_2}(I^2)$, satisfying $Bf|_{G(X,Y)} = f|_{G(X,Y)}$, and Bf is explicitly given by

$$Bf = \mathcal{L}_x f + \mathcal{L}_y f - \mathcal{L}_x \mathcal{L}_y f, \tag{3.2}$$

where $\mathcal{L}_x$ and $\mathcal{L}_y$ are the Lagrange interpolation operators with respect to variables x and y, defined by the interpolation points X and Y, respectively. In other words, if $\{l_\mu\}_{\mu=1}^{m_1}$ and $\{\bar{l}_\nu\}_{\nu=1}^{m_2}$ are the Lagrange fundamental polynomials for π_{m_1-1} and π_{m_2-1}, respectively, defined by $l_\mu(x_j) = \delta_{\mu j}$ $(j = 1, \ldots, m_1)$ and $\bar{l}_\nu(y_k) = \delta_{\nu k}$ $(k = 1, \ldots, m_2)$, δ_{ij} being the Kronecker symbol, then

$$Bf(x, y) = \sum_{\mu=1}^{m_1} l_\mu(x) f(x_\mu, y) + \sum_{\nu=1}^{m_2} \bar{l}_\nu(y) f(x, y_\nu) - \sum_{\mu=1}^{m_1} \sum_{\nu=1}^{m_2} l_\mu(x) \bar{l}_\nu(y) f(x_\mu, y_\nu).$$

For any $r, s \in \mathbb{N}$ satisfying $r \leq m_1$, $s \leq m_2$ and for $f \in C^{r,s}(I^2)$ two iterated applications of the Peano representation theorem to $f - Bf = (Id - \mathcal{L}_x)(Id - \mathcal{L}_y)f$ (Id being the identity operator) implies, for $(x, y) \in I^2$,

$$f(x, y) - Bf(x, y) = \iint_{I^2} \mathcal{K}_r(x, t) \overline{\mathcal{K}}_s(y, \tau) f^{(r,s)}(t, \tau) \, dt \, d\tau, \tag{3.3}$$

where

$$\mathcal{K}_r(x, t) = \frac{1}{(r-1)!}[(x - t)_+^{r-1} - \sum_{\mu=1}^{m_1} l_\mu(x)(x_\mu - t)_+^{r-1}],$$

$$\overline{\mathcal{K}}_s(y, \tau) = \frac{1}{(s-1)!}[(y - \tau)_+^{s-1} - \sum_{\nu=1}^{m_2} \bar{l}_\nu(y)(y_\nu - \tau)_+^{s-1}].$$

The replacement of $L[f]$ by $C[f] := L[Bf]$ yields a *blending cubature formula*

$$C[f] := \sum_{\mu=1}^{m_1} b_\mu \ell[f(x_\mu, \cdot)] + \sum_{\nu=1}^{m_2} \bar{b}_\nu \ell[f(\cdot, y_\nu)] - \sum_{\mu=1}^{m_1} \sum_{\nu=1}^{m_2} b_\mu \bar{b}_\nu f(x_\mu, y_\nu), \tag{3.4}$$

where $Q'[g] = \sum_{\mu=1}^{m_1} b_\mu g(x_\mu)$ and $Q''[g] = \sum_{\nu=1}^{m_2} \bar{b}_\nu g(y_\nu)$ are the interpolatory quadrature formulae generated by $\mathcal{L}_x$ and $\mathcal{L}_y$. Integrating (3.3) we find

$$E[C; f] := L[f] - C[f] = \iint_{I^2} K_r(Q'; t) K_s(Q''; \tau) f^{(r,s)}(t, \tau) \, dt d\tau, \qquad (3.5)$$

and, because of the separated variables, the Peano constants $c_{(r,s),p}(C)$ in the sharp error estimate

$$|E[C; f]| \le c_{(r,s),p}(C) \|f^{(r,s)}\|_p \qquad (3.6)$$

are expressed as

$$c_{(r,s),p}(C) = c_{r,p}(Q') c_{s,p}(Q'')$$

(here and in what follows, $\|h\|_p$ is the $L_p[I^2]$- or $L_p[I]$-norm depending on whether h is bivariate or univariate function). By analogy with the univariate case, we define the Sobolev classes

$$W^{r,s}(I^2) := \{f : I^2 \to \mathbb{R} : \|f^{(r,s)}\|_p < \infty\}.$$

A disadvantage of the blending cubature formula $C[f]$ is that it involves $m_1 + m_2$ univariate integrals, which, in general, are not easy to be calculated exactly. On the other hand, a better approximation to $L[f]$ is usually obtained either by increasing m_1 and m_2 or by use of compound blending cubature formulae based on $C[f]$; both approaches lead to an increasing of the number of line integrals involved.

In our construction, the number of line integrals stays fixed (as well as the integrals themselves), while better approximations to $L[f]$ are achieved by increasing the number of point evaluations.

Let $C_n[f]$ be any cubature formula for $L[f]$ of the form

$$C_n[f] = \sum_{i=0}^{n} \sum_{j=0}^{n} a_{ij} f(t_i, \tau_j), \quad (t_i, \tau_j) \in I^2.$$

We approximate $L[f - Bf] \approx C_n[f - Bf]$ (instead of $L[f] \approx L[Bf]$). This results in a cubature formula $S_n[f]$,

$$L[f] \approx C_n[f] + L[Bf] - C_n[Bf] =: S_n[f]. \qquad (3.7)$$

From (3.3) and (3.5) we find

$$L[f] - S_n[f] = \iint_{I^2} K_{r,s}(S_n; t, \tau) f^{(r,s)}(t, \tau) \, dt d\tau, \qquad (3.8)$$

$$K_{r,s}(S_n; t, \tau) = K_r(Q'; t) K_s(Q''; \tau) - \sum_{i=0}^{n} \sum_{j=0}^{n} a_{ij} \mathcal{K}_r(t_i, t) \overline{\mathcal{K}}_s(\tau_j, \tau). \qquad (3.9)$$

Application of Hölder's inequality to the right–hand side of (3.8) yields sharp error estimates of the type (3.6) for $|E[S_n, f]|$. As in the univariate case, we may have definite cubature formulae.

Definition 1. The cubature formula S_n is called positive (negative) definite of order (r, s) if $K_{r,s}(S_n; t, \tau) \geq 0$ (≤ 0) on I^2.

Next we consider in more detail the case when C_n is a product cubature formula obtained from a quadrature formula $Q_n[g] = \sum_{i=0}^{n} a_i g(t_i)$, i.e., $a_{ij} = a_i a_j$ and $\tau_j = t_j$ for $i, j = 0, \ldots, n$.

Theorem 3.1. *Let in the cubature formula S_n in (3.7) C_n be the product cubature formula, associated with a quadrature formula $Q_n[g] = \sum_{i=0}^{n} a_i g(t_i)$, and let $ADP(Q_n) \geq \max\{ADP(Q'), ADP(Q'')\}$, where Q' and Q'' are the interpolatory quadrature formulae generated by $\mathcal{L}_x$ and $\mathcal{L}_y$, respectively.*

(i) *For $(t, \tau) \in I^2$, the Peano kernel $K_{r,s}(S_n; t, \tau)$ has the following representations:*

$$K_{r,s}(S_n; t, \tau) = K_r(Q'; t) K_s(Q''; \tau) - Q_n[\mathcal{K}_r(\cdot\ ; t)] Q_n[\overline{\mathcal{K}}_s(\cdot\ , \tau)], \qquad (3.10)$$

$$K_{r,s}(S_n; t, \tau) = K_r(Q'; t) K_s(Q_n; \tau) + K_s(Q_n; t) Q_n[\overline{\mathcal{K}}_s(\cdot\ , \tau)], \qquad (3.11)$$

$$K_{r,s}(S_n; t, \tau) = K_s(Q''; \tau) K_r(Q_n; t) + K_s(Q_n; \tau) Q_n[\mathcal{K}_r(\cdot\ , t)], \qquad (3.12)$$

$$\begin{aligned} K_{r,s}(S_n; t, \tau) = \ & K_r(Q'; t) K_s(Q_n; \tau) + K_r(Q_n; t) K_s(Q''; \tau) \\ & - K_r(Q_n; t) K_s(Q_n; \tau). \end{aligned} \qquad (3.13)$$

(ii) *If Q', Q'' and Q_n are symmetric quadrature formulae, then S_n is a symmetric cubature formula. Moreover, $K_{r,s}(S_n; t, \tau)$ is an even or odd function with respect to t (and with respect to τ) depending on whether r (resp. s) is even or odd.*

(iii) *The cubature formula S_n can be represented as*

$$S_n[f] = C_n[f] + Q' \left[R[Q_n; f((\cdot)_{Q'}, (\cdot)_R)] \right] + Q'' \left[R[Q_n; f((\cdot)_R, (\cdot)_{Q''})] \right]. \qquad (3.14)$$

Proof. Since C_n is a product cubature formula, (3.10) is simply (3.9). To prove (3.11) and (3.12), we use that $K_r(Q'; t) = \ell[\mathcal{K}_r(\cdot, t)]$, $K_s(Q''; \tau) = \ell[\overline{\mathcal{K}}_s(\cdot, \tau)]$. Adding and subtracting $K_r(Q'; t) Q_n[\overline{\mathcal{K}}_s(\cdot\ , \tau)]$ in (3.10) and taking into account that $R[Q_n; \overline{\mathcal{K}}_s(\cdot\ , \tau)] = K_s(Q_n; \tau)$ and $R[Q_n; \mathcal{K}_r(\cdot\ , t)] = K_r(Q_n; t)$, we get

(3.11). Equation (3.12) is proved similarly. Now (3.13) follows from (3.11) and from the exactness of Q_n for l_μ, since

$$Q_n[\overline{K}_s(\,\cdot\,,\tau)] = Q_n\left[\frac{(\cdot - \tau)_+^{s-1}}{(s-1)!}\right] - Q'\left[\frac{(\cdot - \tau)_+^{s-1}}{(s-1)!}\right] = K_s(Q';\tau) - K_s(Q_n;\tau).$$

Next, we prove (iii). Using again that $Q_n[l_\mu] = b_\mu$, $Q_n[\bar{l}_\nu] = \bar{b}_\nu$, we find from the explicit form of Bf that

$$C_n[Bf] = \sum_{\mu=1}^{m_1} b_\mu Q_n[f(x_\mu,\cdot)] + \sum_{\nu=1}^{m_2} \bar{b}_\nu Q_n[f(\cdot,y_\nu)] - \sum_{\mu=1}^{m_1}\sum_{\nu=1}^{m_2} b_\mu \bar{b}_\nu f(x_\mu, y_\nu).$$

This last formula and equations (3.7) and (3.4) complete the proof of (iii).

Part (ii) of Theorem 3.1 holds in the more general case when C_n is an arbitrary symmetric cubature formula, not necessarily of product type. It is an easy exercise to prove that S_n is symmetric; then the second claim in (ii) is deduced from (3.8): e.g., replacement of $f(x, y)$ by $f(-x, y)$ does not change the left–hand side, while in the right–hand side $K_{r,s}(S_n; t, \tau)$ is replaced by $(-1)^r K_{r,s}(S_n; -t, \tau)$. We omit the details. $\square$

An immediate consequence of Theorem 3.1 is

Corollary 3.2. *Under the assumptions of Theorem* 3.1, *we have*

(i) $c_{(r,s),p}(S_n) \leq c_{r,p}(Q')c_{s,p}(Q_n) + c_{s,p}(Q'')c_{r,p}(Q_n) + c_{r,p}(Q_n)c_{s,p}(Q_n),$

(ii) $|E[C_n; f]| \leq c_{(r,s),p_1}(S_n)\|f^{(r,s)}\|_{p_1}$

$$+\|Q'\|_\infty c_{\mu,p_2}(Q_n)\|f^{(0,\mu)}\|_{p_2} + \|Q''\|_\infty c_{\nu,p_3}(Q_n)\|f^{(\nu,0)}\|_{p_3}$$

for any $\mu \leq m_1$, $\nu \leq m_2$ *and* $p_i \geq 1$, $i = 1, 2, 3$.

4 Cubature Formulae Using Integrals over ∂I^2

Throughout the next two sections, we shall consider cubature formulae

$$S_n[f] = L[Bf] + C_n[f] - C_n[Bf], \tag{4.1}$$

where C_n is the product cubature formula associated with a quadrature formula Q_n, using equidistant nodes, i.e.,

$$C_n[f] = \sum_{i=0}^{n}\sum_{j=0}^{n} a_i a_j f(t_i, \tau_j) \quad \text{with } t_i = \tau_i = -1 + 2i/n \ \ (i = 0, \ldots, n). \tag{4.2}$$

In this section, Bf interpolates f on ∂I^2, the boundary of I^2. This means that $m_1 = m_2 = 2$, $X = Y = \{-1, 1\}$ and $Q'[g] \equiv Q''[g] = g(-1) + g(1)$.

Definition 2. The cubature formula (4.1) is said to be of quasi–Chebyshev type and is denoted by $S_n^{C,\beta}$, if $C_n[f] = \beta^2 \sum_{i=0}^{n} \sum_{j=0}^{n} f(t_i, \tau_j)$.

Theorem 4.1. *Among all cubature formulae $S_n^{C,\beta}$ of quasi-Chebyshev type, the best one in $W_p^{1,1}(I^2)$, $p = 1, 2$, is obtained for $\beta_* = 2/n$. Moreover,*

$$|E[S_n^{B,\beta_*}]| \le \frac{2}{n}\left(1 - \frac{1}{2n}\right)\|f^{(1,1)}\|_1, \tag{4.3}$$

$$|E[S_n^{B,\beta_*}]| \le \frac{2\sqrt{2}}{3n}\left(1 - \frac{1}{2n^2}\right)^{1/2}\|f^{(1,1)}\|_2, \tag{4.4}$$

$$|E[S_n^{B,\beta_*}]| < \frac{40}{27n}\left(1 + \frac{27}{40n} - \frac{2}{5n^2}\right)\|f^{(1,1)}\|_\infty. \tag{4.5}$$

Remark 2. The meaning of "best" is that no $S_n^{C,\beta}$ exists with Peano constant $c_{(1,1),p}(S_n^{C,\beta})$, $p = 1, 2$, smaller than those appearing in the right–hand sides of (4.3) and (4.4). Though the error constant in (4.5) is not exact, its leading term $40/27n$ is sharp.

Proof. For $(t, \tau) \in I^2$ we have

$$K_{1,1}(S_n^{C,\beta}; t, \tau) = t\tau - \beta^2 \sum_{i=0}^{n} \mathcal{K}_1(t_i, t) \sum_{j=0}^{n} \mathcal{K}_1(\tau_j, \tau), \tag{4.6}$$

where $\mathcal{K}_1(x, t) = (x - t)_+^0 - (1 + x)/2$. We need to compare the L_q–norms of $K_{1,1}(S_n^{C,\beta}; \cdot, \cdot)$ for $q = \infty$ and 2. For $\mu, \nu \in \{0, \ldots, n - 1\}$, we set

$$\Delta_{\mu\nu} := \{(t, \tau) : t_\mu < t < t_{\mu+1}, \tau_\nu < \tau < \tau_{\nu+1}\}. \tag{4.7}$$

If $(t, \tau) \in \Delta_{\mu,\nu}$, then we find from (4.6)

$$K_{1,1}\left(S_n^{C,\beta}; t_\mu + \frac{1+u}{n}, \tau_\nu + \frac{1+v}{n}\right) = \frac{4 - n^2\beta^2}{4n^2}(n - 2\mu - 1)(n - 2\nu - 1)$$

$$+ \frac{1}{n^2}[uv - (n - 2\nu - 1)u - (n - 2\mu - 1)v]. \tag{4.8}$$

It follows from (4.8) that if $|\beta| < 2/n$, then the maximum of $K_{1,1}(S_n^{C,\beta}; \cdot, \cdot)$ is attained at $(t_0 + 0, \tau_0 + 0)$ and $(t_n - 0, \tau_n - 0)$, while if $|\beta| > 2/n$, then the maximum of $K_{1,1}(S_n^{C,\beta}; \cdot, \cdot)$ is attained at $(t_0 + 0, \tau_n - 0)$ and $(t_n - 0, \tau_0 + 0)$; in both cases

$$\max_{(t,\tau) \in I^2} |K_{1,1}(S_n^{C,\beta}; t, \tau)| > \frac{2n-1}{n^2}.$$

If $|\beta| = 2/n$, then the latter inequality becomes equality, which proves (4.3).

Next, we see from (4.8) that

$$\iint_{\Delta_{\mu\nu}} [K_{1,1}(S_n^{C,\beta}; t, \tau)]^2 dt\, d\tau = \frac{1}{n^2} \iint_{I^2} \left[\frac{4 - n^2\beta^2}{4n^2}(n - 2\mu - 1)(n - 2\nu - 1) \right.$$

$$\left. + \frac{1}{n^2}[uv - (n - 2\nu - 1)u - (n - 2\mu - 1)v] \right]^2 du\, dv.$$

The two summands appearing in the second integral are orthogonal on I^2, hence

$$\iint_{\Delta_{\mu\nu}} [K_{1,1}(S_n^{C,\beta}; t, \tau)]^2 dt\, d\tau \geq \iint_{\Delta_{\mu\nu}} [K_{1,1}(S_n^{C,\beta*}; t, \tau)]^2 dt\, d\tau,$$

and consequently $\|K_{1,1}(S_n^{C,\beta}; \cdot, \cdot)\|_2 \geq \|K_{1,1}(S_n^{C,\beta*}; \cdot, \cdot)\|_2$ with equality if and only if $|\beta| = |\beta*|$. The calculation of $\|K_{1,1}(S_n^{C,\beta*}; \cdot, \cdot)\|_2$ is straightforward and therefore is omitted.

For the proof of (4.5) we estimate $\iint_{\Delta_{\mu\nu}} |K_{1,1}(S_n^{C,\beta*}; t, \tau)|\, dt\, d\tau$ by

$$\frac{1}{n^4} \iint_{I^2} [|(n - 2\nu - 1)u + (n - 2\mu - 1)v| \pm |uv|]\, du\, dv,$$

and calculate this integral with the help of a simple lemma.

Lemma 4.2. *If $a \geq b > 0$, then $\iint_{I^2} |au + bv|\, du\, dv = 2a + 2b^2/(3a)$.*

The details are left to the reader. $\qquad\square$

Now consider general cubature formulae (4.1) with C_n being the product cubature formula (4.2) generated by a quadrature formula Q_n with $\text{ADP}(Q_n) \geq 1$. In view of Theorem 3.1, we have

$$K_{2,2}(S_n; t, \tau) = \frac{\tau^2 - 1}{2} K_2(Q_n; t) + K_2(Q_n; \tau)Q_n[K_2(\cdot, t)], \qquad (4.9)$$

$$K_{2,2}(S_n; t, \tau) = \frac{t^2 - 1}{2} K_2(Q_n; \tau) + \frac{\tau^2 - 1}{2} K_2(Q_n; t) \qquad (4.10)$$
$$- K_2(Q_n; t)K_2(Q_n; \tau),$$

where $\mathcal{K}_2(x,t) = (x-t)_+ - (1+x)(1-t)/2$.

Theorem 4.3. (i) *If Q_n is a negative definite quadrature formula of order 2, then S_n is a positive definite cubature formula of order (2,2).*

(ii) *If Q_n is a positive definite quadrature formula of order 2 and Q_n has only positive coefficients, then S_n is a negative definite cubature formula of order (2,2).*

Proof. Part (ii) follows immediately from (4.10), and part (i) follows from (4.9), since $\mathcal{K}_2(x,t) \leq 0$ on I^2.   □

If we choose Q_n to be the trapezium quadrature formula (2.6), the resulting cubature formula is ($f_{ij} := f(t_i, \tau_j)$)

$$S_n[f] = \ell[f(t_0, \cdot)] + \ell[f(t_n, \cdot)] + \ell[f(\cdot, \tau_0)] + \ell[f(\cdot, \tau_n)] + \frac{4}{n^2} \sum_{i=1}^{n-1} \sum_{j=1}^{n-1} f_{ij}$$

$$- \frac{2n-1}{n^2}[f_{00} + f_{0n} + f_{n0} + f_{nn}] - \frac{2n-2}{n^2} \sum_{k=1}^{n-1} [f_{0k} + f_{nk} + f_{k0} + f_{kn}].$$

$$\tag{4.11}$$

Actually, S_n^{C,β_*} is identical with (4.11), due to the fact that $(f - Bf)_{|\partial I^2} \equiv 0$. Thus, Theorem 4.1 reveals some extremal properties of the cubature formula (4.11). Further properties are given in the next theorem.

Theorem 4.4. *The cubature formula (4.11) is positive definite of order (2.2). Moreover, if $f^{(2,2)} \geq 0$ on I^2, then*

$$L[f] \geq S_{2n}[f] \geq S_n[f]. \tag{4.12}$$

The following error estimates for S_n hold true:

$$|E[S_n; f]| < \frac{1}{2n^2} \|f^{(2,2)}\|_1, \tag{4.13}$$

$$|E[S_n; f]| \leq \frac{2}{45n^2} \left(122 - \frac{100}{n^2} + \frac{14}{n^4}\right)^{1/2} \|f^{(2,2)}\|_2, \tag{4.14}$$

$$|E[S_n; f]| \leq \frac{8}{9n^2} \left(1 - \frac{1}{2n^2}\right) \|f^{(2,2)}\|_\infty. \tag{4.15}$$

Proof. The positive definiteness of S_n follows from Theorem 4.3 (i). The second inequality in (4.12) follows from the fact that $K_{2,2}(S_n; t, \tau) \geq K_{2,2}(S_{2n}; t, \tau)$

on I^2. In order to see that, we assume, e.g., that $(t, \tau) \in \Delta_{2\mu, 2\nu}$, where $0 \le \mu, \nu \le n - 1$ and $t_i = \tau_i := -1 + i/n$. Then

$$
\begin{aligned}
K_{2,2}(S_n; t, \tau) = & \; [(t^2 - 1)(\tau - \tau_{2\nu})(\tau - \tau_{2\nu+2}) + (\tau^2 - 1)(t - t_{2\mu})(t - t_{2\mu+2}) \\
& - (t - t_{2\mu})(t - t_{2\mu+2})(\tau - \tau_{2\nu})(\tau - \tau_{2\nu+2})] / 4,
\end{aligned}
$$

$$
\begin{aligned}
K_{2,2}(S_{2n}; t, \tau) = & \; [(t^2 - 1)(\tau - \tau_{2\nu})(\tau - \tau_{2\nu+1}) + (\tau^2 - 1)(t - t_{2\mu})(t - t_{2\mu+1}) \\
& - (t - t_{2\mu})(t - t_{2\mu+1})(\tau - \tau_{2\nu})(\tau - \tau_{2\nu+1})] / 4.
\end{aligned}
$$

Set

$$
\psi(u, v) := K_{2,2}(S_n; t_{2\mu} + u/n, \tau_{2\nu} + v/n) - K_{2,2}(S_{2n}; t_{2\mu} + u/n, \tau_{2\nu} + v/n)
$$

where $u, v \in [0, 1]$, then

$$
\begin{aligned}
\psi(u, v) = & \; \frac{1}{4n^4} \{4\nu(n - \nu)u + 4\mu(n - \mu)v - [2(2\mu - n) + 2(2\nu - n) + 3]uv\} \\
\ge & \; \frac{uv}{2n^4} [2(\mu + 1)(n - \mu) + 2(\nu + 1)(n - \nu) - 2n - 2] \ge 0.
\end{aligned}
$$

The cases when (t, τ) belongs to $\Delta_{2\mu+1, 2\nu}$, $\Delta_{2\mu, 2\nu+1}$ and $\Delta_{2\mu+1, 2\nu+1}$ are treated similarly and lead to the same conclusion.

Application of the triangle inequality to (4.10) proves (4.13):

$$
\|K_{2,2}(S_n; \cdot, \cdot)\|_\infty \le \|K_2(Q_n^{tr}; \cdot)\|_\infty = \frac{1}{2n^2}.
$$

It can easily be seen that this estimate is asymptotically sharp, e.g., by calculating $K_{2,2}(S_n; 0, 0)$ for n odd.

The proof of (4.14) and (4.15) is done by calculation of $\|K_{2,2}(S_n; \cdot, \cdot)\|_q$ for $q = 2$ and $q = 1$. Using (4.10) and the fact that $K_{2,2}(S_n; \cdot, \cdot) \ge 0$, we prove (4.15):

$$
\|K_{2,2}(S_n; \cdot, \cdot)\|_1 = \frac{4}{3} c_{2,\infty}(Q_n^{tr}) - [c_{2,\infty}(Q_n^{tr})]^2 = \frac{8}{9n^2} \left(1 - \frac{1}{2n^2}\right).
$$

We need the quantities $d_n^{tr} := \|((\cdot)^2 - 1)K_2(Q_n^{tr}; \cdot)\|_1$ and $c_{2,2}(Q_n^{tr})$ to determine $\|K_{2,2}(S_n; \cdot, \cdot)\|_2$. A short calculation shows that

$$
d_n^{tr} = \frac{1}{2} \sum_{\nu=0}^{n-1} \int_{t_\nu}^{t_{\nu+1}} (t^2 - 1)(t - t_\nu)(t - t_{\nu+1}) \, dt = \frac{4}{9n^2} + \frac{4}{45n^4},
$$

while $[c_{2,2}(Q_n^{tr})]^2 = 4/15n^2$, and from (4.10) we get

$$\|K_{2,2}(S_n;\cdot,\cdot)\|_2^2 = \frac{8}{15}[c_{2,2}(Q_n^{tr})]^2 + [c_{2,2}(Q_n^{tr})]^4 + \frac{1}{2}[d_n^{tr}]^2 - 2d_n^{tr}[c_{2,2}(Q_n^{tr})]^2$$

$$= \frac{4}{45^2 n^4}\left(122 - \frac{100}{n^2} + \frac{14}{n^4}\right),$$

i.e. Theorem 4.4 is proved.   □

If (4.1) is constructed with C_n as in (4.2), generated by the quadrature formula Q_n^+ defined by (2.5), then S_n has the following (less elegant) form:

$$S_n[f] = \ell[f(t_0,\cdot)] + \ell[f(t_n,\cdot)] + \ell[f(\cdot,\tau_0)] + \ell[f(\cdot,\tau_n)] + \frac{4}{n^2}\sum_{i=3}^{n-3}\sum_{j=3}^{n-3}f_{ij}$$

$$-\frac{15}{4n}[f_{01} + f_{10} + f_{0,n-1} + f_{n-1,0} + f_{1n} + f_{n1} + f_{n-1,n} + f_{n,n-1}]$$

$$-\frac{5}{4n}[f_{02} + f_{20} + f_{0,n-2} + f_{n-2,0} + f_{2n} + f_{n2} + f_{n-2,n} + f_{n,n-2}]$$

$$+\frac{75}{16n^2}[f_{12} + f_{21} + f_{1,n-2} + f_{n-2,1} + f_{2,n-1} + f_{n-1,2} + f_{n-2,n-1} + f_{n-1,n-2}]$$

$$+\frac{15}{2n^2}\sum_{k=3}^{n-3}[f_{1k} + f_{k1} + f_{n-1,k} + f_{k,n-1}] + \frac{5}{2n^2}\sum_{k=3}^{n-3}[f_{2k} + f_{k2} + f_{n-2,k} + f_{k,n-2}]$$

$$-\frac{2}{n}\sum_{k=3}^{n-3}[f_{0k} + f_{nk} + f_{k0} + f_{kn}] + \frac{25}{16n^2}[f_{22} + f_{2,n-2} + f_{n-2,2} + f_{n-2,n-2}]$$

$$+\frac{225}{16n^2}[f_{11} + f_{1,n-1} + f_{n-1,1} + f_{n-1,n-1}].$$

$$(4.16)$$

Theorem 4.5. *The cubature formula* (4.16) *is negative definite of order* (2.2), *i.e., if* $f^{(2,2)} \geq 0$ *on* I^2, *then* $L[f] \leq S_n[f]$. *Moreover,*

$$|E[S_n; f]| \leq \frac{1}{2n^2}\left(1 + \frac{1}{2n^2}\right)\|f^{(2,2)}\|_1, \qquad (4.17)$$

$$|E[S_n; f]| < \frac{1}{n^2}\left(\frac{158}{2025} + \frac{56}{15n} + \frac{410}{81n^2} - \frac{80}{27n^3} + \frac{7004}{27n^4}\right)^{1/2}\|f^{(2,2)}\|_2, \quad (4.18)$$

$$|E[S_n; f]| \leq \frac{4}{9n^2}\left(1 + \frac{15}{n}\right)\left(1 + \frac{1}{4n^2} + \frac{15}{4n^3}\right)\|f^{(2,2)}\|_\infty. \qquad (4.19)$$

Proof. The negative definiteness of (4.16) follows from Theorem 4.3 (ii). For the proof of the inequalities (4.17), (4.18) and (4.19) we have to find $\|K_{2,2}(S_n;\cdot,\cdot)\|_q$ for $q = \infty, 2$ and 1, and to this end we use the representation (4.10). It is seen from Theorem 2.1 (i) that $c_{2,\infty}(Q_n^+) = 1/(3n^2) + 5/n^3$, while straightforward calculation shows that $[c_{2,2}(Q_n^+)]^2 = 1/(10n^4) + 7/n^5$, and

$$d_n^+ := \int_{-1}^{1} (1 - t^2) K_2(Q_n; t)\, dt = \frac{2}{9n^2} + \frac{1016}{45n^4} - \frac{82}{3n^5}.$$

We have $\|K_{2,2}(S_n;\cdot,\cdot)\|_1 = (4/3)c_{2,\infty}(Q_n^+) + [c_{2,\infty}(Q_n^+)]^2$, and replacement of $c_{2,\infty}(Q_n^+)$ proves (4.19). The exact constant in (4.18) is slightly sharper: we have

$$\|K_{2,2}(S_n;\cdot,\cdot)\|_2^2 = \frac{8}{15}[c_{2,2}(Q_n^+)]^2 + [c_{2,2}(Q_n^+)]^4 + \frac{1}{2}[d_n^+]^2 - 2d_n^+[c_{2,2}(Q_n^+)]^2$$

$$= \frac{158}{2025n^4} + \frac{56}{15n^5} + \frac{410}{81n^6} - \frac{80}{27n^7} + \frac{2101169}{8100n^8} - \frac{41189}{135n^9} + \frac{359}{9n^{10}}.$$

Though $\|K_2(Q_n^+;\cdot)\|_\infty = K_2(Q_n^+;\pm t_2) = 2/n^2$, $K_{2,2}(S_n;t,\tau)$ attains its maximum norm at or near to the origin, due to the fact that $|t^2 - 1|$ and $|\tau^2 - 1|$ are small near ∂I^2 (note that $n \geq 5$). Since $\|K_2(Q_n^+;\cdot)\|_{C[t_2,t_{n-2}]} = 1/2n^2$, we deduce that $\|K_{2,2}(S_n;\cdot,\cdot)\|_\infty \leq \|K_2(Q_n^+;\cdot)\|_{C[t_2,t_{n-2}]} + \left[\|K_2(Q_n^+;\cdot)\|_{C[t_2,t_{n-2}]}\right]^2$, which proves (4.17). $\qquad\square$

5 Formulae Using Integrals Over Coordinate Axes

Consider the blending functions $B_0 f \in B^{1,1}(I^2)$ and $Bf \in B^{2,2}(I^2)$, given by

$$B_0 f(x, y) = f(0, y) + f(x, 0) - f(0, 0), \tag{5.1}$$

and

$$Bf(x, y) = f(0, y) + xf^{(1,0)}(0, y) + f(x, 0) + yf^{(0,1)}(x, 0)$$
$$- f(0, 0) - xf^{(1,0)}(0, 0) - yf^{(0,1)}(0, 0) - xyf^{(1,1)}(0, 0), \tag{5.2}$$

respectively. Clearly, $B_0 f$ and Bf interpolate f on the same blending grid G, generated by $X = Y = \{0\}$, but Bf interpolates f in the Hermite sense (see [1]). However, the quadrature formulae Q' and Q'' associated with (5.1) and (5.2) are identical and coincide with the midpoint rule $Q[g] = 2g(0)$. Furthermore, it is easy to see that if C_n is the product cubature formula generated by a quadrature formula Q_n with $\text{ADP}(Q_n) \geq 1$, then $B_0 f$ and Bf generate the same cubature formula S_n (see also Theorem 3.1 (iii)). Therefore, S_n does not involve derivatives of f. For $(t, \tau) \in I^2$ Theorem 3.1 yields

$$
\begin{aligned}
K_{2,2}(S_n; t, \tau) &= \frac{(1 - |t|)^2}{2} K_2(Q_n, \tau) + \frac{(1 - |\tau|)^2}{2} K_2(Q_n, t) \\
&\quad - K_2(Q_n, t) K_2(Q_n, \tau),
\end{aligned}
\tag{5.3}
$$

$$
K_{2,2}(S_n; t, \tau) = \frac{(1 - |t|)^2}{2} K_2(Q_n, \tau) + K_2(Q_n, t) Q_n[K_2(\cdot, \tau)],
\tag{5.4}
$$

where $K_2(x, t) := (x - t)_+ - (-t)_+ - x(-t)_+^0$. Since $K_2(x, t) \geq 0$ on I^2, the above representations of $K_{2,2}(S_n; t, \tau)$ immediately imply

Theorem 5.1. (i) *If Q_n is a negative definite quadrature formula of order 2, then S_n is a negative definite cubature formula of order (2,2);*

(ii) *If Q_n is a positive definite quadrature formula of order 2 and Q_n has only positive coefficients, then S_n is a positive definite cubature formula of order (2,2).*

We now apply Theorem 5.1 to obtain definite cubature formulae of order (2,2), which involve line integrals over the coordinate axes. If Q_n is the trapezium quadrature formula, then we get the cubature formula

$$
\begin{aligned}
S_n[f] =\ & 2\ell[f(t_{\frac{n}{2}}, \cdot)] + 2\ell[f(\cdot, \tau_{\frac{n}{2}})] + \frac{1}{n^2}[f_{00} + f_{0n} + f_{n0} + f_{nn}] \\
& + \frac{2}{n^2} \sum_{k=1}^{n-1} [f_{0k} + f_{nk} + f_{k0} + f_{kn}] + \frac{4}{n^2} \sum_{i=1}^{n-1} \sum_{j=1}^{n-1} f_{ij} \\
& - \frac{2}{n}[f_{\frac{n}{2}, 0} + f_{\frac{n}{2}, n} + f_{0, \frac{n}{2}} + f_{n, \frac{n}{2}}] - \frac{4}{n} \sum_{k=1}^{n-1} [f_{\frac{n}{2}, k} + f_{k, \frac{n}{2}}]
\end{aligned}
\tag{5.5}
$$

Theorem 5.2. *The cubature formula (5.5) is negative definite of order (2.2). Moreover, if $f^{(2,2)} \geq 0$ on I^2, then*

$$
L[f] \leq S_{2n}[f] \leq S_n[f].
\tag{5.6}
$$

The following error estimates for S_n hold true:

$$
|E[S_n; f]| \leq \frac{1}{2n^2} \left(1 + \frac{1}{2n^2}\right) \|f^{(2,2)}\|_1,
\tag{5.7}
$$

$$|E[S_n; f]| \leq \frac{1}{45n^2} \left(158 + \frac{275}{n^2} + \frac{1873}{8n^4}\right)^{1/2} \|f^{(2,2)}\|_2, \tag{5.8}$$

$$|E[S_n; f]| \leq \frac{4}{9n^2} \left(1 + \frac{1}{n^2}\right) \|f^{(2,2)}\|_\infty. \tag{5.9}$$

Theorem 5.3. *Let, for $n \geq 5$, $S_n[f] = C_n[f] + L[Bf] - C_n[Bf]$ with Bf given in (5.2) and C_n being the product cubature formula generated by the quadrature formula Q_n^+ in (2.5). Then S_n is a positive definite cubature formula of order (2.2), i.e., if $f^{(2,2)} \geq 0$ on I^2, then $L[f] \geq S_n[f]$. Moreover,*

$$|E[S_n; f]| < \frac{1}{2n^2} \|f^{(2,2)}\|_1, \tag{5.10}$$

$$|E[S_n; f]| < \frac{1}{45n^2} \left(53 + \frac{2835}{n} - \frac{25}{n^2} + \frac{3211}{n^3}\right)^{1/2} \|f^{(2,2)}\|_2, \tag{5.11}$$

$$|E[S_n; f]| \leq \frac{2}{9n^2} \left(1 + \frac{15}{n}\right) \left(1 - \frac{1}{2n^2} - \frac{15}{2n^3}\right) \|f^{(2,2)}\|_\infty. \tag{5.12}$$

We omit the proof of Theorems 5.2 and 5.3, since it goes along the same lines as the proof of Theorems 4.4 and 4.5, with (4.10) replaced by (5.3). Note that all the estimates are sharp or asymptotically sharp.

6   Concluding Remarks

1. The definite cubature formulae considered in Sections 4 and 5 do not differ essentially in the error constants. However, they are of different kind of definiteness and/or use different line integrals. It is well–known that definite quadrature formulae are far from being optimal in the Sobolev classes of functions. The same is true in the bivariate case, too. For instance, if Q_n is the asymptotically optimal quadrature formula (2.4) in $W_\infty^2(I)$, then the application of Corollary 3.2 (i) yields crude, but better error bounds for the resulting cubature formulae:

$$|E[S_n; f]| \leq \frac{1}{3n^2} \left(1 + \frac{533}{384n}\right) \left(1 + \frac{3}{16n^2} + \frac{533}{2048n^2}\right) \|f^{(2,2)}\|_\infty \tag{6.1}$$

for the cubature formula which uses line integrals over ∂I^2, and

$$|E[S_n; f]| \leq \frac{1}{6n^2} \left(1 + \frac{533}{384n}\right) \left(1 + \frac{3}{8n^2} + \frac{533}{1024n^2}\right) \|f^{(2,2)}\|_\infty \tag{6.2}$$

for the cubature formula which uses line integrals over the segments of coordinate axes contained in I^2. Similarly, better estimates in $W_2^{2,2}(I^2)$ are obtained if the quadrature formula (2.3) is used.

2. For those who prefer (or are only able) to use product cubature formulae $C_n[f]$ rather than their "exotic" counterpart $S_n[f] = C_n[f] + L[Bf] - C_n[Bf]$, Corollary 3.2 (ii) furnishes error estimates in terms of only three derivatives of the integrand. Though these estimates are not sharp, they seem more accessible than the error bounds given in [12, Chapter 5], which involve norms of all the partial derivatives of the integrand up to a certain order. The contribution of the additional term $L[Bf] - C_n[Bf]$ to the error is indeterminate: it may reduce or increase the error. We show that, under some additional assumptions for the integrand, the definite cubature formulae presented in Theorems 4.3, 4.4, 5.2 and 5.3 are superior to the product cubature formulae therein.

Theorem 6.1. *Let $f \in W_1^{2,2}(I^2)$, and let $f^{(2,2)}$ do not change its sign on I^2.*

(i) *If $f^{(2,0)}(x, \pm 1)$ and $f^{(0,2)}(\pm 1, y)$ have the same constant sign for $(x, y) \in I^2$, opposite to $\operatorname{sign} f^{(2,2)}$, then for the cubature formulae S_n in Theorems 4.4 and 4.5 and the product cubature formulae therein we have*

$$|E[C_n; f]| \geq |E[S_n; f]|. \tag{6.3}$$

(ii) *If $f^{(2,0)}(x, 0)$ and $f^{(0,2)}(0, y)$ have the same constant sign as $f^{(2,2)}$ for $(x, y) \in I^2$, then for the cubature formulae in Theorems 5.2 and 5.3 and the product cubature formulae therein (6.3) holds true.*

Proof. From (3.14) we find for the cubature formulae in Theorems 4.4 and 4.5

$$\begin{aligned} E[C_n; f] = \ & E[S_n; f] + R[Q_n; f(-1, \cdot)] + R[Q_n; f(1, \cdot)] \\ & + R[Q_n; f(\cdot, -1)] + R[Q_n; f(\cdot, 1)], \end{aligned}$$

and for the cubature formulae in Theorems 5.2 and 5.3

$$E[C_n; f] = E[S_n; f] + 2R[Q_n; f(0, \cdot)] + 2R[Q_n; f(\cdot, 0)].$$

Since S_n and Q_n are definite of opposite kind in the first case, and of the same kind in the second case, we see that all terms in the right hand sides have the same sign, and hence (6.3) holds.   □

Actually, the only property used in the proof is the definiteness of Q_n (S_n is definite as a consequence). Therefore, such a comparison result follows for any definite cubature formula S_n obtained by the scheme proposed in the last two sections, and the product cubature formula C_n involved in S_n.

3. The analogue of Theorem 4.1 for best cubature formulae of quasi–Chebyshev type associated with the blending interpolant (5.1) is given below without proof. Note that these cubature formulae are of inferior quality, as they are essentially designed for functions from $W^{(1,1)}(I^2)$, i.e., of very low smoothness.

Theorem 6.2. (a) *Among all cubature formulae* $S_n^{C,\beta}$, *derived from the blending interpolant* (5.1), *the best one in* $W_1^{1,1}(I^2)$ *is obtained for* $\beta_*^2 = 4(n^2 - 2n + 2)/n^4$, *if* n *is even, and for* $\beta_*^2 = 4(n-1)^2/[n^2(n^2+3)]$, *if* $n \geq 5$ *is odd. Moreover,*

$$|E[S_n^{B,\beta_*}]| \leq \frac{2}{n}\left(1 - \frac{1}{n}\right)\|f^{(1,1)}\|_1, \; \textit{if } n \textit{ is even,}$$

and

$$|E[S_n^{B,\beta_*}]| \leq \frac{2(n-1)^2(n+1)}{n^2(n^2+3)}\|f^{(1,1)}\|_1, \; \textit{if } n \geq 5 \textit{ is odd.}$$

(b) *Among all cubature formulae* $S_n^{C,\beta}$, *derived from the blending interpolant* (5.1), *the best one in* $W_2^{1,1}(I^2)$ *is obtained for* $\beta_* = (2n-1)/(n(n+1))$, *if* n *is even, and for* $\beta_* = (2n+1)/(n^2+2n+3)$, *if* n *is odd. Moreover,*

$$|E[S_n^{B,\beta_*}]| \leq \frac{[(7n-2)(8n^3+8n^2-7n+2)]^{1/2}}{6n^2(n+1)}\|f^{(1,1)}\|_2, \; \textit{if } n \textit{ is even,}$$

and

$$|E[S_n^{B,\beta_*}]| \leq \frac{[(7n-1)(8n^3+16n^2+17n+1)]^{1/2}}{6n(n^2+2n+3)}\|f^{(1,1)}\|_2, \; \textit{if } n \textit{ is odd.}$$

Acknowledgment. This work was supported by the Bulgarian Ministry of Education and Science through Grant MM–802.

References

[1]   B. D. Bojanov, D. P. Dryanov, W. Haußmann, G. P. Nikolov: *Best one–sided L^1–approximation by blending functions*, in: Advances in Multivariate Approximation, W. Haußmann, K. Jetter, M. Reimer (eds.), Wiley–VCH, Berlin–Weinheim–New York–Chichester–Brisbane–Singapore–Toronto 1999, pp. 86–106.

[2]   H. Braß: *Quadraturverfahren*, Vandenhoeck & Ruprecht, Göttingen 1977.

[3]   H. Braß, K.-J. Förster: *On the application of the Peano repre-sentation of linear functionals in numerical analysis*, in: Recent Progress in Inequalities, G. Milovanovic (ed.), Kluwer, Dordrecht 1998.

[4]   P. J. Davis, P. Rabinovitz: *Methods of Numerical Integration*, Academic Press, New York–San Francisco–London 1975.

[5]   K.-J. Förster: *Exit criteria and monotonicity in compound quad-ratures*, Numer. Math. **66** (1993), 321–327.

[6]   K.-J. Förster: *A survey of stopping rules in quadrature based on Peano kernel methods*, Suppl. Rend. Circ. Mat. Palermo, Ser. II, **33** (1993), 311–330.

[7]   K.-J. Förster, P. Köhler, G. P. Nikolov: *Monotonicity and stop-ping rules for compound Gauss–type quadrature formulae*, East J. Approx. **4** (1998), 55–74.

[8]   V. I. Krylov: *Approximate Calculation of Integrals*, Nauka, Moscow 1967 (in Russian). English translation of the first edi-tion: MacMillan, New York–London 1962.

[9]   I. Mysovskikh: *Interpolatory Cubature Formulae*, Nauka, Moscow 1982.

[10]   A. Sard: *Linear Approximation*, Math. Surveys and Monographs **9**, Amer. Math. Soc., Providence, R. I., 1982 (Second printing).

[11]   G. Schmeißer: *Optimale Quadraturformeln mit semidefiniten Ker-nen*, Numer. Math. **20** (1972), 32–53.

[12]   A. H. Stroud: *Approximate Calculation of Multiple Integrals*, Prentice–Hall, Englewood Cliffs, N. J., 1971.

Address:

Vesselin Gushev, Geno Nikolov
Department of Mathematics
University of Sofia
5 James Bourchier Blvd.
BG–1164 Sofia
Bulgaria

International Series of Numerical Mathematics
Vol. 137, ©2001 Birkhäuser Verlag Basel/Switzerland

On an Extremal Problem Originating in Questions of Unconditional Convergence

Hermann König

1 The Problem and the Grothendieck Inequality

Problem. Let $n \in \mathbb{N}$ and define $K : \mathbb{R}^n \times \mathbb{R}^n \to \mathbb{R}$ by

$$K(x,y) := \sin(<x,y>)\exp(-(\|x\|^2 + \|y\|^2)/2); \; x,y \in \mathbb{R}^n$$

where $\| \cdot \|$ is the euclidean norm and $< \cdot, \cdot >$ is the standard scalar product on $\mathbb{R}^n$. Define $T_K : L_\infty(\mathbb{R}^n) \to L_1(\mathbb{R}^n)$ by

$$T_K f(x) = \int_{\mathbb{R}^n} K(x,y)f(y)dy, \; x \in \mathbb{R}^n$$

i.e. T_K is the exponentially weighted odd part of the Fourier transform on $\mathbb{R}^n$. Is it true that the operator norm of T_K is attained on functions like $\operatorname{sgn} x_1$, i.e. is

$$\|T_K : L_\infty(\mathbb{R}^n) \to L_1(\mathbb{R}^n)\|_{Op} = \|T_k(\operatorname{sgn} x_1)\|_{L_1(\mathbb{R}^n)}, \tag{1.1}$$

for any dimension n?

Since K is odd, any extremal function needs to be odd. Clearly $\operatorname{sgn}(x_1)$ where $x = (x_1, \ldots, x_n)$ might be replaced in (1.1) – if true – by any function $\operatorname{sgn}(<x,e>)$ where $e \in S^{n-1} \subset \mathbb{R}^n$ is a fixed direction. Formula (1.1) holds for $n = 1$; this was shown by N. Tomczak–Jaegermann and myself (unpublished). It is unknown if $n \geq 2$.

The problem is motivated by the question of determining the best constant K_G in the so–called Grothendieck inequality which was fundamental in Grothendieck's theory of topological tensor products of Banach spaces [1] and which states (in a somewhat different language):

Grothendieck inequality. There is a constant $K_G < \infty$ such that for any $m \in \mathbb{N}$ and any matrix $K = (k_{ij})_{i,j \leq m}$ satisfying

$$\|K : l_\infty \to l_1\| \leq 1, \quad \text{i.e.} \quad \left| \sum_{i,j} k_{ij} s_i t_j \right| \leq \sup_i |s_i| \sup_j |t_j|; \quad s_i, t_j \in \mathbb{R}$$

we have for any Hilbert space and any set of vectors $(x_i), (y_i) \subset H$ $(i, j \leq m)$

$$\left| \sum_{i,j} k_{ij} < x_i, y_j > \right| \leq K_G \sup_i \|x_i\| \sup_j \|y_j\|.$$

Let K_G denote the best (smallest) possible such constant. Grothendieck's proof yielded $K_G \leq \sinh(\pi/2) \simeq 2.3$. Grothendieck also showed that $K_G \geq \pi/2$. The inequality is crucial in studying the relation between absolute and unconditional convergence of series in Banach spaces. In infinite dimensional Banach spaces, unconditional convergence is a strictly weaker notion than absolute convergence of series. The Grothendieck inequality yields e.g. that, in a converse sense, any operator from a space $L_1(\mu)$ into a space $L_2(\nu)$ transforms unconditionally convergent into absolutely convergent series. Further, it has many applications in the theory of absolutely p–summing operators and the constant K_G appears in many estimates, cf. e.g. [4]. Hence various authors have attempted – so far in vain – to determine the precise value of K_G. Krivine [3] showed that

$$K_G \leq \pi / \left(2 \ln(1 + \sqrt{2}) \right) \simeq 1.782,$$

and this value is conjectured to be optimal. Krivine's proof uses positive definite functions on spheres S^{n-1}. It is straightforward to extend Grothendieck's inequality to integral operators instead of matrices with the same value of K_G:

If (Ω, μ) is a measure space and $K \in L_1((\Omega, \mu)^2)$ and $\varphi, \psi \in L_\infty((\Omega, \mu); H)$, then the following inequality holds

$$\left| \int_\Omega \int_\Omega K(x, y) < \varphi(x), \psi(y) > d\mu(y) d\mu(x) \right| \leq$$

$$\leq K_G \|T_K : L_\infty(\Omega, \mu) \to L_1(\Omega, \mu)\| \operatorname*{ess\,sup}_{x \in \Omega} \|\varphi(x)\| \operatorname*{ess\,sup}_{y \in \Omega} \|\psi(y)\| \quad (1.2)$$

2 Spherical Integration

Analyzing Krivine's proof, U. Haagerup found good reasons that the kernel $K(x,y) = \sin(<x,y>)\exp(-(\|x\|^2 + \|y\|^2)/2)$ on $\Omega = \mathbb{R}^n$ should be extremal for (1.2) in an asymptotic sense, i.e. for $n \to \infty$, **if $K_G = \pi/(2\ln(1+\sqrt{2}))$** holds true.

A calculation by N. Tomczak–Jaegermann and myself shows that, indeed, Krivine's value is the Grothendieck constant if (1.1) were true:

Proposition. *If the conjecture (1.1) is true for (a subsequence of integers) $n \to \infty$, then $K_G = \pi/(2\ln(1+\sqrt{2}))$.*

Proof. Take $\Omega = \mathbb{R}^n$ and let K be the kernel in the problem. Let $\varphi : \mathbb{R}^n \to S^{n-1}, \varphi(x) = x/\|x\|$. Assuming (1.1), (1.2) yields

$$K_G \geq \frac{\int_{\mathbb{R}^n}\int_{\mathbb{R}^n} K(x,y) < \frac{x}{\|x\|}, \frac{y}{\|y\|} > dx\, dy}{\|T_K : L_\infty(\mathbb{R}^n) \to L_1(\mathbb{R}^n)\|}$$

$$= \frac{\int_{\mathbb{R}^n}\int_{\mathbb{R}^n} K(x,y) < \frac{x}{\|x\|}, \frac{y}{\|y\|} > dx\, dy}{\int_{\mathbb{R}^n} |\int_{\mathbb{R}^n} K(x,y)\mathrm{sgn}(y_1)dy|\, dx}$$

$$= \frac{\int_{\mathbb{R}^n}\int_{\mathbb{R}^n} K(x,y) < \frac{x}{\|x\|}, \frac{y}{\|y\|} > dx\, dy}{\int_{\mathbb{R}^n}\int_{\mathbb{R}^n} K(x,y)\, \mathrm{sgn}\, y_1\, \mathrm{sgn}\, x_1\, dy\, dx} =: \frac{I_n}{J_n}. \tag{2.1}$$

We show that I_n/J_n is increasing with $\lim_n I_n/J_n = \dfrac{\pi}{2\ln(1+\sqrt{2})}$: Introducing polar coordinates, the integral in the numerator of (2.1) equals

$$I_n = \int_0^\infty \int_0^\infty (st)^{n-1} e^{-(s^2+t^2)/2} \times$$

$$\times \left(\int_{S^{n-1}} \int_{S^{n-1}} \sin(st <u,v>) <u,v> du\, dv \right) ds\, dt. \tag{2.2}$$

Spherical integration of the inner integral gives, independently of $u \in S^{n-1}$,

$$\int_{S^{n-1}} \sin(r < u, v >) < u, v > dv = |S^{n-2}| \, 2 \int_0^1 \sin(rt)t(1-t^2)^{(n-3)/2} dt.$$

Letting $t = \cos \theta$ and integrating by parts, the latter integral from 0 to 1 equals

$$\int_0^{\pi/2} \sin(r \cos \theta)(\cos \theta)(\sin \theta)^{n-2} d\theta = \frac{r}{n-1} \int_0^{\pi/2} \cos(r \cos \theta)(\sin \theta)^n d\theta$$

$$= \frac{r}{n-1} \frac{\sqrt{\pi}}{2} \Gamma\left(\frac{n+1}{2}\right) \left(\frac{2}{r}\right)^{n/2} J_{n/2}(r)$$

where $J_{n/2}$ is the Bessel function of order $n/2$. This, (2.2) and

$$|S^{n-1}| \, |S^{n-2}| = \frac{4\pi^{n-1/2}}{\Gamma\left(\frac{n}{2}\right) \Gamma\left(\frac{n-1}{2}\right)}$$

imply that

$$I_n = c_n \int_0^\infty \int_0^\infty (st)^{n/2} J_{n/2}(st) e^{-(s^2+t^2)/2} ds \, dt$$

$$= c_n \int_0^\infty \left(\int_0^\infty J_{n/2}(u) u^{n/2} e^{-u^2/(2t^2)} du \right) e^{-t^2/2} \frac{dt}{t}$$

with $c_n = 2^{n/2+1}\pi^n / \Gamma\left(\frac{n}{2}\right)$. By formula (6.631/1) of [2], the inner integral equals the modified hypergeometric function

$$d_n t^{n+1} \, {}_1F_1\left(\frac{n+1}{2}; \frac{n}{2} + 1; -t^2/2\right), d_n := \frac{\Gamma\left(\frac{n+1}{2}\right)}{\sqrt{2}\,\Gamma\left(\frac{n}{2}+1\right)}.$$

We conclude

$$I_n = c_n d_n \int_0^\infty {}_1F_1\left(\frac{n+1}{2}; \frac{n}{2} + 1; -t^2/2\right) t^n e^{-t^2/2} dt$$

$$= c_n d_n 2^{(n-1)/2} \int_0^\infty {}_1F_1\left(\frac{n+1}{2}; \frac{n}{2} + 1; -u\right) u^{\frac{n-1}{2}} e^{-u} du.$$

By formula (7.525/1) of [2], the last integral from 0 to ∞ equals

$$\Gamma\left(\frac{n+1}{2}\right)\,_2F_1\left(\frac{n+1}{2},\frac{n+1}{2};\frac{n}{2}+1;-1\right)$$

which, using the transformation formula for hypergeometric functions,

$$_2F_1(\alpha,\beta;\gamma;z) = (1-z)^{\gamma-(\alpha+\beta)}\,_2F_1(\gamma-\alpha,\gamma-\beta;\gamma;z),$$

see (9.131/1) in [2], is the same as

$$\Gamma\left(\frac{n+1}{2}\right)2^{-n/2}\,_2F_1\left(\frac{1}{2},\frac{1}{2};\frac{n}{2}+1;-1\right).$$

This means that

$$I_n = c_n d_n/\sqrt{2}\,\Gamma\left(\frac{n+1}{2}\right)\,_2F_1\left(\frac{1}{2},\frac{1}{2};\frac{n}{2}+1;-1\right)$$

$$= 2^{n/2}\pi^n\frac{\Gamma\left(\frac{n+1}{2}\right)^2}{\Gamma\left(\frac{n}{2}\right)\Gamma\left(\frac{n}{2}+1\right)}\,_2F_1\left(\frac{1}{2},\frac{1}{2};\frac{n}{2}+1;-1\right). \qquad (2.3)$$

To evaluate the integral in the denominator of (2.1), we write $x,y \in \mathbb{R}^n$ as

$$x = (x_1,x'), \quad y = (y_1,y') \in \mathbb{R} \times \mathbb{R}^{n-1},$$
$$<x,y> = x_1 y_1 + <x',y'>.$$

Hence

$$\sin(<x,y>) = \sin(x_1 y_1)\cos(<x',y'>) + \cos(x_1 y_1)\sin(<x',y'>).$$

Since $\text{sgn}(y_1)\cos(x_1 y_1)$ is odd in y_1, the integral

$$\int_{\mathbb{R}} \cos(x_1 y_1)\text{sgn}(y_1)e^{-y_1^2/2}dy_1$$

is 0. This means for the denominator of (2.1)

$$J_n = \left(\int_{\mathbb{R}}\int_{\mathbb{R}} \sin(x_1 y_1)\text{sgn}(x_1 y_1)e^{-(x_1^2+y_1^2)/2}dx_1\,dy_1\right)\cdot$$

$$\cdot\left(\int_{\mathbb{R}^{n-1}}\int_{\mathbb{R}^{n-1}} \cos(<x',y'>)e^{-(\|x'\|^2+\|y'\|^2)/2}dx'\,dy'\right).$$

Since $e^{-\|y'\|^2/2}$ is, up to the factor $(2\pi)^{(n-1)/2}$, a fixed point of the Fourier transform in $\mathbb{R}^n$, the second integral equals

$$(2\pi)^{(n-1)/2} \int_{\mathbb{R}^{n-1}} e^{-\|x'\|^2/2} dx' = 2^{(n-1)/2} \pi^{n-1}.$$

Therefore

$$J_n = 2^{(n+3)/2} \pi^{n-1} \int_0^\infty \int_0^\infty \sin(x_1 y_1) e^{-(x_1^2+y_1^2)/2} dx_1\, dy_1.$$

Using again polar coordinates, $x_1 = r\cos\varphi$, $y_1 = r\sin\varphi$, letting $u = r^2/2$, $\psi = 2\varphi$, one gets, integrating by parts twice

$$
\begin{aligned}
J_n &= 2^{(n+3)/2} \pi^{n-1} \int_0^{\pi/2} \left(\int_0^\infty \sin(u\sin\psi)e^{-u} du \right) d\psi \\
&= 2^{(n+3)/2} \pi^{n-1} \int_0^{\pi/2} \frac{\sin\psi}{1+\sin^2\psi} d\psi.
\end{aligned}
$$

The substitution $v = (\tan(\psi/2))^2$ yields

$$J_n = 2^{(n+5)/2} \pi^{n-1} \int_0^1 \frac{dv}{1+6v+v^2} = 2^{\frac{n}{2}+1} \pi^{n-1} \ln(1+\sqrt{2}). \qquad (2.4)$$

Assuming the conjecture (1.1) to be true, we have (2.1) and hence by (2.3) and (2.4)

$$K_G \geq \frac{I_n}{J_n} = \frac{\pi}{2\ln(1+\sqrt{2})} \left(\frac{\Gamma\left(\frac{n+1}{2}\right)^2}{\Gamma\left(\frac{n}{2}\right)\Gamma\left(\frac{n}{2}+1\right)} \right) {}_2F_1\left(\frac{1}{2},\frac{1}{2};\frac{n}{2}+1;-1\right)$$

which for $n \to \infty$ tends to $\pi/(2\ln(1+\sqrt{2}))$. Thus $K_G = \pi/(2\ln(1+\sqrt{2}))$ would hold, the upper bound being Krivine's.

Note here that

$$f(n) := \frac{\Gamma\left(\frac{n+1}{2}\right)^2}{\Gamma\left(\frac{n}{2}\right)\Gamma\left(\frac{n}{2}+1\right)} \, {}_2F_1\left(\frac{1}{2},\frac{1}{2};\frac{n}{2}+1;-1\right)$$

if of order $1 - \frac{1}{n} + 0\left(\frac{1}{n^2}\right)$ as $n \to \infty$ (and increasing in $n \in \mathbb{N}$).   □

For $n = 1$, (1.1) is true. The difficulty of the problem originates with the fact that the regions of positivity and negativity of $\sin(< x, y >)$ – regions between hyperbolas if $n = 1$ – have a different shape than the grid–structured rectangular regions which are obtained by taking tensor products of extreme points f, g in the unit ball of $L_\infty(\mathbb{R}^n)$, e.g. for $n = 1$ taking f and g of the type

$$f = \sum_{i=0}^{M}(-1)^n \chi_{[c_i,c_{i+1})}, 0 = c_0 < c_1 < c_2 \ldots,$$

in trying to estimate

$$\int_{\mathbb{R}^n}\int_{\mathbb{R}^n} k(x,y)f(x)g(y)dx\,dy$$

to prove (1.1).

References

[1]   A. Grothendieck: *Resumé de la theorie metrique des produits ten-soriels topologiques*, Bol. Soc. Mat. Sao Paolo **8** (1956), 1–79.

[2]   I. Gradhsteyn, I. Ryzhik: *Table of Integrals, Series and Products*, 5th edition, Acad. Press, 1994.

[3]   J. L. Krivine: *Constants de Grothendieck et fonctions de type positif sur les sphères*, Advances in Math. **31** (1979), 16–30.

[4]   J. Lindenstrauss, L. Tzafriri: *Classical Banach spaces I*, Springer, 1977.

Address:

Hermann König
Mathematisches Seminar
Universität Kiel
D–24098 Kiel
Germany

International Series of Numerical Mathematics
Vol. 137, ©2001 Birkhäuser Verlag Basel/Switzerland

Node Insertion and Node Deletion for Radial Basis Functions

Alain Le Méhauté and Yvon Lafranche

Abstract

The present paper is a contribution to data reduction in the multivariate case. In this direction, there are quite a few methods like ours dealing with scattered data points. Besides piecewise polynomial splines, radial basis functions (RBF splines) provide another natural generalization of univariate splines. Considering a set A of distinct scattered data points and a given tolerance ϵ, our aim is to extract a subset $\mathcal{A} \subset A$ such that the RBF spline $\sigma_{\mathcal{A}}$ stays within a tolerance ϵ from the RBF spline σ_A built upon the entire set A of data points. We show that for the usual RBF splines, data reduction can be considered from two different points of view: *a priori* reduction which amounts to adding nodes one after the other, starting from a small subset of A, thus refering to a rough approximation of σ_A, and *a posteriori* reduction, which in contrast consists of deleting nodes one after the other from the ultimate spline σ_A.

1 Introduction

We consider interpolation and approximation of real valued functions defined on a set $\Omega \subset \mathbb{R}^d, d \geq 1$.

The functions are evaluated on a set $X = \{x_1, x_2, \ldots, x_N\}$ of N pairwise distinct points in Ω, usually called the *centers* or the *nodes*, and we are dealing here only with Lagrange interpolation at each of those points, *i.e.*, the evaluation functionals are restricted to $\delta_{x_j} : f \longmapsto f(x_j), j = 1, 2, \ldots, N$. Hermite interpolation or Hermite Birkhoff interpolation can also be investigated [1,7,12].

We want to interpolate by some function in the space

$$V_\varphi = V_{\varphi,X} = \text{span}\{\varphi(\bullet - x_\ell) : \ell = 1, 2, \ldots, N\} \ .$$

Here, φ is a radial function or more precisely, a radially symmetric function $\varphi(x) = g(\|x\|_2^2)$, where $\|x\|_2$ denotes the Euclidian norm of the vector x, and $g : [0, \infty) \mapsto \mathbb{R}$ is a univariate function.

It is usually assumed that φ is continuous, and is the inverse Fourier transform of an $L_1(\mathbb{R}^d)$ function $\hat{\varphi}$, i.e.,

$$\varphi(x) = \frac{1}{(2\pi)^d} \int_{\mathbb{R}^d} \hat{\varphi}(\xi) \, e^{ix\cdot\xi} d\xi \qquad \text{for all } x \in \mathbb{R}^d,$$

with $x \cdot \xi$ the usual dot product in $\mathbb{R}^d$.

Let $S = S[f] \in V_{\varphi,X}$ be the interpolant to f on the set X. Thus

$$S(x) = S[f](x) = \sum_{j=1}^{N} \alpha_j \, \varphi(x - x_j) = \sum_{j=1}^{N} \alpha_j \, g(\|x - x_j\|_2^2) ,$$

and the coefficients α_j, $j = 1, \ldots, N$ are chosen such that to satisfy the interpolation conditions

$$S(x_k) = f(x_k) = f_k , \qquad k = 1, \ldots, N.$$

Consequently, these coefficients are the solution of the linear system

$$\begin{cases} \sum_{j=1}^{N} \alpha_j \, \varphi(x_k - x_j) = f_k , \\ \\ k = 1, \ldots, N, \end{cases}$$

provided that the symmetric matrix (the collocation matrix)

$$A = A_{\varphi,X} = \begin{pmatrix} \varphi(x_1 - x_1) & \cdots & \varphi(x_1 - x_N) \\ \vdots & \cdots & \vdots \\ \varphi(x_N - x_1) & \cdots & \varphi(x_N - x_N) \end{pmatrix}$$

is non–singular.

This is the simplest form of radial basis function interpolation, the more generic form being the following:

Let $P_q[\mathbb{R}^d]$ be the space of d–variate real polynomials of degree $\leq q$. Let $Q = \dim P_q[\mathbb{R}^d] = \binom{q+d}{d}$, and select a basis $\{p_1, p_2, \ldots, p_Q\}$ of $P_q[\mathbb{R}^d]$.

We consider now the interpolation space

$$V_{\varphi,q} = V_{\varphi,q,X} = V_{\varphi,X} \oplus P_q[\mathbb{R}^d].$$

This means that we are looking for interpolating functions of the form

$$s(x) = s[f](x) = \sum_{j=1}^{N} \alpha_j \, \varphi(x - x_j) + \sum_{\ell=1}^{Q} \beta_\ell \, p_\ell(x).$$

The Q additional degrees of freedom are compensated by an additional set of Q equations, corresponding to the fact that we require $V_{\varphi,X}$ to be in a certain sense orthogonal to $P_q[\mathbb{R}^d]$, i.e.,

$$\begin{cases} \sum_{k=1}^{N} \alpha_j \, p_\ell(x_k) = 0, \\ k = 1,\dots,Q. \end{cases}$$

Let P_X be the $(N \times Q)$–matrix

$$P_X = \begin{pmatrix} p_1(x_1),\dots,p_Q(x_1) \\ \dots\dots\dots\dots \\ p_1(x_N),\dots,p_Q(x_N) \end{pmatrix}.$$

The interpolation conditions are still $s(x_k) = f(x_k) = f_k$, $k = 1,\dots,N$, i.e.,

$$\begin{cases} \sum_{j=1}^{N} \alpha_j \, \varphi(x_k - x_j) + \sum_{\ell=1}^{Q} \beta_\ell \, p_\ell(x_k) = f_k\ , \\ k = 1,\dots,N, \end{cases}$$

and the coefficients vectors $\boldsymbol{\alpha} = \{\alpha_1,\alpha_2,\dots,\alpha_N\}$ and $\boldsymbol{\beta} = \{\beta_1,\beta_2,\dots,\beta_Q\}$ are solutions to the linear system

$$\begin{pmatrix} A_{\varphi,X} & P_X \\ P_X^\top & 0 \end{pmatrix} \begin{pmatrix} \boldsymbol{\alpha} \\ \boldsymbol{\beta} \end{pmatrix} = \begin{pmatrix} \boldsymbol{f} \\ 0 \end{pmatrix}$$

where $\boldsymbol{f}^\top = (f_1, f_2,\dots,f_N)$.

Such a system is uniquely solvable whenever φ is conditionally positive definite of order q and X is $P_q[\mathbb{R}^d]$–unisolvent, meaning that:

Definition 1.1. *A function $\phi : \mathbb{R}^d \longmapsto \mathbb{R}$ with $\phi(-x) = \phi(x)$, is said to be conditionally positive definite of order q on $\mathbb{R}^d$ if, for any set $X = \{x_1, x_2,\dots,x_N\}$ of N distinct points in $\mathbb{R}^d$ the following holds true: For all non–zero vectors $\boldsymbol{\alpha} = (\alpha_1, \alpha_2,\dots,\alpha_N) \in \mathbb{R}^d$, satisfying the moment equations*

$$\sum_{j=1}^{N} \alpha_j \, p_\ell(x_j) = 0\ , \quad \ell = 1,2,\dots,Q,$$

one has

$$\sum_{j,k=1}^{N} \alpha_j \, \alpha_k \, \phi(x_j - x_k) > 0\ .$$

Definition 1.2. *A set X of points in $\mathbb{R}^d$ is called $P_q[\mathbb{R}^d]$–unisolvent, if the only polynomial $p \in P_q[\mathbb{R}^d]$ which vanishes on X is identically zero.*

For the matrices A and P, these requirements are usually stated in the following form:

rank$(P_X) = Q \leq N$, *and there exists a positive constant* $\lambda = \lambda_N$ *such that, for all non–zero* $\alpha \in \mathbb{R}^N$ *satisfying* $P_X^\top \alpha = 0$, *one has* $\alpha^\top A\,\alpha \geq \lambda\,\|\alpha\|_2^2$.

If q is negative, then $P_q[\mathbb{R}^d]$ is empty, and thus P_X and β do not show up in the equations. In this case, $\|A^{-1}\|_2 \leq \lambda^{-1}$ holds in the spectral norm, and the quantity λ^{-1} controls the sensitivity of the solution vector α with respect to variations in the data vector f.

The most commonly used among the radial basis functions are the following (as usual, $r = \|x\|$):

Multiquadrics: $\varphi(r) = (c^2 + r^2)^{\frac{\beta}{2}}$ with $\beta > -d, \beta \notin 2\mathbb{Z}$, and $2q > \beta$.

Polyharmonic splines:

    Thin plate splines: $\varphi(r) = (-1)^{\frac{\beta}{2}+1} r^\beta \log r$, with $\beta \in 2\mathbb{N}$ and $2q > \beta$.

    Pseudo–polynomial splines: $\varphi(r) = r^\beta$, with $\beta > 0, \beta \notin 2\mathbb{Z}$ and $2q > \beta$.

Shifted thin plate splines: $\varphi(r) = (-1)^{\frac{\beta}{2}+1} (c^2 + r^2)^\beta \log(c^2 + r^2)$,

$$\text{with } \beta \in 2\mathbb{N} \text{ and } 2q > \beta.$$

Gaussians: $\varphi(r) = e^{-\beta r^2}$ with $\beta > 0$ and $q \geq 0$.

However, when dealing with an interpolation process to approximate given data, one would often appreciate to decrease the number of the knots $\{x_j\}_{j=1}^N$ that are initially used, in order to facilitate the evaluation of $s[f]$ without changing s beyond a given uniform tolerance. More generally, one wants to eliminate nodes from general approximants (not necessarily interpolants) in order to facilitate their computation. One reason for this may be that s has to be evaluated often on a computer, once an initial interpolant or approximant has been found.

We therefore study in this paper the effect of removing one center (or node) at a time from a given interpolant, or, what is more or less equivalent, the effect of inserting one center at a time to a given interpolant.

2 Adding or Removing Centers

The aim of this section is to study theoretically the effect of inserting one center to a given interpolant or removing one center at a time from a given interpolant. We make no claim as to the optimality of the bounds, since the work on this subject is still in progress.

2.1. A Priori Bounds

Given $m < N$ points $\{x_j : j = 1, 2, \ldots, m\} \subset X$, and $x_{new} \in X$, $x_{new} \notin \{x_j\}_{j=1}^m$, we compare the radial function s_{old} interpolating f at $\{x_j : j = 1, 2, \ldots, m\}$, i.e., $s_{old}(x_j) = f(x_j) = f_j$ for $j = 1, 2, \ldots, m$, with the new interpolant s_{new} that interpolates the data $\{f_j : j = 1, 2, \ldots, m\}$ and f_{new} at $\{x_j : j = 1, 2, \ldots, m\}$ and x_{new}, respectively. This comparison is performed in the uniform norm.

We restrict ourselves to multiquadric radial functions $\varphi(r) = (c^2 + r^2)^{\frac{\beta}{2}}$ with $\beta \notin 2\mathbb{Z}$ and $-d < \beta < 2$.

Let Ω be a compact domain in $\mathbb{R}^d$ which contains all the centers x_j, $j = 0, \ldots, m$ including the new one $x_0 = x_{new}$. We put

$$h = \max_{1 \leq j \leq m} \min_{1 \leq k \leq m} \| x_j - x_k \|$$

and $\|g\|_\infty = \|g\|_{\infty,\Omega} = \sup_{x \in \Omega} |g(x)|$. Finally, A is the matrix

$$A = [\varphi(\|x_j - x_k\|)]_{j,k=0}^m ,$$

and $\lambda(A)$ denotes the smallest eigenvalue, in modulus, of A.

Theorem 2.1. [5, 6] *Given any arbitrary positive integer M, there exists a constant C, depending on M but independent of h, x_{new} and f_{new}, such that*

$$\| s_{old} - s_{new} \|_{\infty,\Omega} \leq C \, (\lambda(A))^{-\frac{1}{2}} \, |f_{new} - s_{old}(x_{new})| \, h^M .$$

More general results are available. They apply to all polyharmonic splines $\varphi(r)$ and to shifted versions $\varphi(\sqrt{r^2 + c^2})$ thereof. It should be noted that this covers most of the commonly used radial functions which are globally supported.

2.2. A Posteriori Bounds

Given points $\{x_j\}_{j=1}^m$ and x_{new}, all in $\mathbb{R}^d$ and distinct, we want to compare the radial basis function s_{old} which interpolates $f' = \{f(x_j)\}_{j=1}^m$ at $\{x_j\}_{j=1}^m$ with the interpolant s_{new} that interpolates $f_{new} = f(x_{new})$ as well.

We restrict ourself in the beginning of this section to radial functions of the form

$$\varphi(r) = (r^2 + c^2)^{\frac{\beta}{2}}, \quad -d < \beta < 2,$$

for which interpolants with no additional polynomial term exist.

Theorem 2.2. *Let $M = \lfloor \beta \rfloor + d + 1$. Under the given conditions, we have the estimate*

$$\|s_{old} - s_{new}\|_{\infty,\Omega} \leq C \ \|\boldsymbol{a}'\|_2^{-2}\Lambda(A') \ |f_{new} - s_{old}(x_{new})| \, h^M$$

where $h = \max_{1 \leq j \leq m} \min_{1 \leq k \leq m} \|x_j - x_k\|_2$ is the (Euclidian) spacing of the centers, A' is the matrix $A' = \left[\varphi(\|x_j - x_k\|_2)\right]_{j,k=1}^m$, $\Lambda(A')$ denotes its largest eigenvalue in modulus, $\boldsymbol{a}'$ is the vector $\left\{\varphi(\|x_j - x_{new}\|_2)\right\}_{j=1}^m$, and C is a positive constant independent on h, x_{new} and f_{new} .

Remark: We note that for the radial functions which are under investigation here we have $\|\boldsymbol{a}'\|_2^{-2} \leq \frac{1}{m} \times \min_{1 \leq j \leq m} \varphi(\|x_{new} - x_j\|_2)^{-2}$ and $\Lambda(A') \leq m \times \max_{1 \leq j,k \leq m} \varphi(\|x_j - x_k\|_2)$. In the very particular case $d = 1$ and $x_{new} = 0 < x_1 < x_2 < \cdots < x_m \leq 1$, this gives $\|\boldsymbol{a}'\|_2^{-2} \ \Lambda(A') \leq \frac{\varphi(1)}{\varphi(x_1)^2}$.

Same kind of results occur when dealing with radial basis functions of the form

$$\varphi(r) = \varphi_0(r) = \begin{cases} r^{2M-d}\ln r, & \text{for d even, } M > \frac{d}{2} \\ r^{2M-d}, & \text{for d odd, } M > \frac{d}{2} \end{cases}$$

or of a shifted version thereof, $\varphi_c(r) = \varphi_0(\sqrt{r^2 + c^2})$. (Even more general formulations exist, see [1, 7], but we do not claim here for more generality). From these results we quote

Theorem 2.3. *Under the stated conditions, we have the estimate*

$$\|s_{old} - s_{new}\|_{\infty,\Omega} \leq C \ \|\boldsymbol{a}'\|_2^{-2}\Lambda(A') \ |f_{new} - s_{old}(x_{new})| \, h^{\frac{2M}{\delta}}$$

where $h = \max_{1 \leq j \leq m} \min_{1 \leq k \leq m} \|x_j - x_k\|_2$ *is the (Euclidian) spacing of the centers,* A' *is the matrix* $A' = \left[\varphi(\|x_j - x_k\|_2)\right]_{j,k=1}^{m}$, $\Lambda(A')$ *denotes its largest eigenvalue in modulus, and* $\boldsymbol{a}'$ *is the vector* $\left\{\varphi(\|x_j - x_{new}\|_2)\right\}_{j=1}^{m}$. *Moreover,* $\delta = 1$ *if* $\varphi = \varphi_c$ *with* $c > 0$, *and* $\delta = 1 + \frac{1}{2}$ *if* $\varphi = \varphi_0$, *and* C *is a positive constant independent on* h, x_{new} *and* f_{new}.

Remark: We note that for the polyharmonic radial functions we have the same estimates as in the remark above.

3 Node Deletion

The main idea in knot removal techniques is – as often with basic ideas – an obvious one :

Given a tolerance ϵ, *check if the point* x_{new} *is such that* $\|s_{old} - s_{new}\|$ *(or some equivalent term) is less than* ϵ *or not .*

If yes, *then remove* x_{new} *from the data list, because it is not significant as compared with the other points already considered;*

If no, *then keep* x_{new} *within the data and check the next point in the list.*

Unfortunately, the results of the previous section are not constructive and for practical purposes, we use the following technique which was taylor–made for multiquadrics. To this end, for a function

$$\sigma(\cdot) = \sum_{i=1}^{N} \alpha_i \phi(\| \cdot - x_i\|^2)$$

we define the discrete p–norms

$$\|\sigma\|_{d,p} = \|\alpha\|_p = (\sum_{i=1}^{N} |\alpha_i|^p)^{1/p} , \quad 1 \leq p < \infty ,$$

and

$$\|\sigma\|_{d,\infty} = \|\alpha\|_\infty = \max_{1 \leq i \leq N} |\alpha_i| .$$

Theorem 3.1. *For the set of radial function approximants* σ *spanned by multiquadrics, the norms* $\|\cdot\|_{d,p}$ *and* $\|\cdot\|_{d,\infty}$ *are equivalent to the uniform norm* $\|\cdot\|_{\infty,\Omega}$.

Proof: Let us recall the usual norms for a matrix $A = [a_{i,j}]$ of order N:

$$\|A\|_1 = \sup_{v \neq 0} \frac{\|Av\|_1}{\|v\|_1} = \max_{1 \leq j \leq N} \sum_{i=1}^{N} |a_{i,j}|,$$

$$\|A\|_2 = \sup_{v \neq 0} \frac{\|Av\|_2}{\|v\|_2} = \sqrt{\rho(A^\top A)} = \|A^\top\|_2,$$

$$\|A\|_\infty = \sup_{v \neq 0} \frac{\|Av\|_\infty}{\|v\|_\infty} = \max_{1 \leq i \leq N} \sum_{j=1}^{N} |a_{i,j}|.$$

From the properties of norms of matrices and vectors,

$$\frac{1}{\sqrt{N}} \| A\alpha \|_2 \leq \| A\alpha \|_\infty$$

and

$$\frac{1}{\sqrt{N}} \| \alpha \|_2 = \frac{1}{\sqrt{N}} \| A A^{-1} \alpha \|_2 \leq \frac{1}{\sqrt{N}} \|A^{-1}\|_2 \| A\alpha \|_2,$$

whence

$$\frac{1}{\sqrt{N} \|A^{-1}\|_2} \| \alpha \|_2 \leq \frac{1}{\sqrt{N}} \| A\alpha \|_2 \leq \| A\alpha \|_\infty.$$

But when interpolating with multiquadrics we have

$$\| A\alpha \|_\infty = \max_{1 \leq i \leq N} \left| \sum_{j=1}^{N} \alpha_i\, \varphi(\|x_i - x_j\|_2) \right| \leq \sup_{x \in \Omega} \left| \sum_{j=1}^{N} \alpha_i\, \varphi(\|x - x_j\|_2) \right|$$

which is in short

$$\| A\alpha \|_\infty \leq \| s \|_{\infty,\Omega}.$$

Now

$$\| s \|_{\infty,\Omega} \leq \|\alpha\|_1 \max_{1 \leq i \leq N} \sup_{x \in \Omega} \left\{ (\|x - x_i\|_2^2 + c^2)^{\frac{1}{2}} \right\},$$

and denoting by $C(\Omega)$ the constant

$$C(\Omega) = \max_{1 \leq i \leq N} \sup_{x \in \Omega} \left\{ (\|x - x_i\|_2^2 + c^2)^{\frac{1}{2}} \right\} \leq (\mathrm{diam}(\Omega)^2 + c^2)^{\frac{1}{2}}$$

it is obvious that the norms are equivalent:

$$\frac{1}{\sqrt{N} \|A^{-1}\|_2} \| s \|_{d,2} \leq \|s\|_{\infty,\Omega} \leq \sqrt{N}\, C(\Omega) \| s \|_{d,2}. \qquad \square$$

Now, let s be the multiquadric interpolating all the data, and let $s' = s_{new}$ be the one interpolating all the data except in one point, say, x_{i_0}. Let us write

$$s(x) = \sum_{i=1}^{N} \alpha_i \, \varphi(\|x - x_i\|) \text{ and}$$

$$s'(x) = \sum_{\substack{i=1 \\ i \neq i_0}}^{N} \alpha_i' \, \varphi(\|x - x_i\|) = \sum_{i=1}^{N} \alpha_i' \, \varphi(\|x - x_i\|)$$

where $\alpha_{i_0}' = 0$. Since $\|s - s'\|_{\infty,\Omega} \leq C(\Omega)\sqrt{N}\,\|s - s'\|_{d,2}$ and $\|s\|_{\infty,\Omega} \geq \frac{1}{\sqrt{N}\,\|A^{-1}\|_2}\,\|s\|_{d,2}$, we obtain the following upper bound for the relative error:

$$\frac{\|s - s'\|_{\infty,\Omega}}{\|s\|_{\infty,\Omega}} \leq C(\Omega)N\|A^{-1}\|_2 \frac{\|s - s'\|_{d,2}}{\|s\|_{d,2}}\,.$$

Here, A is the collocation matrix for the N centers associated with s, while $C(\Omega)$ is a geometric constant related to the diameter of Ω and the parameter c involved in the definition of the multiquadric. Also, by definition,

$$\|s - s'\|_{d,2} = \left(\sum_{i=1}^{N} (\alpha_i - \alpha_i')^2\right)^{\frac{1}{2}}.$$

Let $\boldsymbol{a} = (\alpha_1, \ldots, \alpha_{i_0-1}, \alpha_{i_0+1}, \ldots, \alpha_N)$ and $\boldsymbol{a}' = (\alpha_1', \ldots, \alpha_{i_0-1}', \alpha_{i_0+1}', \ldots, \alpha_N')$. We write the interpolation conditions $s(x_k) = f_k$, $k = 1, \ldots, N$, $k \neq i_0$, as

$$\begin{cases} \sum_{\substack{i=1 \\ i \neq i_0}}^{N} \alpha_i \, \varphi(\|x_k - x_i\|) = f_k - \alpha_{i_0} \, \varphi(\|x_k - x_{i_0}\|) \\ \qquad \text{for } k = 1, 2, \ldots, N, k \neq i_0\,. \end{cases}$$

Let A' be the collocation matrix associated with s'. It is obtained from A by deleting the row and the column associated with i_0.

We have

$$A'(\boldsymbol{a} - \boldsymbol{a}') = -\alpha_{i_0} \begin{pmatrix} \varphi(\|x_1 - x_{i_0}\|) \\ \cdots\cdots \\ \varphi(\|x_N - x_{i_0}\|) \end{pmatrix},$$

and therefore

$$\|\boldsymbol{a} - \boldsymbol{a}'\|_2 \leq |\alpha_{i_0}| \|(A')^{-1}\|_2 \Big\{(N-1)c^2 + \sum_{\substack{i=1 \\ i \neq i_0}}^{N} \|x_i - x_{i_0}\|^2\Big\}$$

$$\leq |\alpha_{i_0}| \|(A')^{-1}\|_2 \Big\{(N-1)c^2 + D^2\Big\},$$

where $D = \max_{1 \leq i,j \leq N} \|x_i - x_k\|$. Thus $\|\alpha - \alpha'\|_2 \leq \left\{ |\alpha_{i_0}|^2 + \|a - a'\|_2^2 \right\}^{\frac{1}{2}}$ provides the upper bound

$$\|s - s'\|_{d,2} \leq |\alpha_{i_0}| \{1 + \|(A')^{-1}\|_2 (N-1)(c^2 + D^2)\}^{1/2}.$$

Now, from [15], we have that

$$\|(A')^{-1}\|_2 \leq 5.95 \frac{1}{p_N} \exp\left(\frac{3}{p_N}\right),$$

with $p_N = q_N^2(1 + (1 + q_N^2)^{1/2})^{-1}$ and q_N the minimal separation distance between the centers, i.e., $q_N = \min_{1 \leq i,j \leq N} \|x_i - x_k\|$. We finally obtain an upper bound of the form

$$\frac{\|\sigma - \sigma'\|_{\infty,\Omega}}{\|\sigma\|_{\infty,\Omega}} \leq K \frac{|\alpha_{i_0}|}{\|\alpha\|_2}$$

with the constant K explicitly given by

$$K = 5.95 C(\Omega) N \frac{1}{p_N} \exp\left(\frac{3}{p_N}\right) \left\{ 1 + \frac{1}{p_N} 5.95(N-1)(c^2 + D^2) \exp\left(\frac{3}{p_N}\right) \right\}^{1/2}.$$

This result is more suitable for practical knot removal techniques when using multiquadrics, because it implies that for any i_0 such that

$$\frac{|\alpha_{i_0}|}{\|\alpha\|_2} \leq \frac{\epsilon}{K},$$

the relative error $\|s - s'\|_\infty / \|s\|_\infty$ for s is less than the given tolerance ϵ.

Now, the usual techniques [11,13,14] for knot removal can be invoked, and we arrive at the following

Algorithm :

- **compute**

  ▷ compute K

  ▷ compute the coefficients α_i of the spline

- **evaluate**

  ▷ for $j = 1, 2, \ldots, N$ compute $w_j = K \dfrac{|\alpha_j|}{\|\alpha\|_2}$

- **rank**

  ▷ rearrange the x_j in increasing order of w_j

- **eliminate**

  ▷ eliminate those x_j such that $w_j < \epsilon$

4 Node Adding and Node Insertion

The main idea in knot adding techniques is also an obvious one :

Given a tolerance ϵ, check if the point x_{new} is such that $\|s_{old} - s_{new}\|$ (or some equivalent term) is less than ϵ or not.

If yes, *then* add x_{new} *to the data list, because this point is significant as compared to the others already considered.*

If no, *keep* x_{new} *out of the data and check for the next point in the list.*

Although being apparent, we would like to point to the following. One has to iterate the process, since a point which is not taken in a first stage can become significant after introduction of some other points. This is because the splines under consideration are globally supported. Of course, this remark holds for node deletion, too.

Unfortunately again, the results of the second section are too theoretical and not constructive enough, and for practical purposes we use the following technique. We consider the construction of the RBF–spline function as an iterative process.

At the beginning, we have a set $X = \{x_1, x_2, \ldots, x_N\}$ of N distinct centers, and we are going to extract from X a subset $X' = \{x'_1, x'_2, \ldots, x'_n\}$ with $n \leq N$ by ignoring the non–significant points. Let us assume that s_k, that is the spline built up from a set X'_k of k data points, is known, i.e.,

$$s_k(x) = \sum_{i=1}^{k} \lambda_{i,k}\, \varphi(\|x - x'_i\|) + \sum_{\ell=1}^{Q} \mu_{\ell,k}\, p_\ell(x)$$

was obtained by solving a linear system $A_k z_k = b_k$ with the matrix A_k built from the centers $X'_k = \{x'_1, x'_2, \ldots, x'_k\}$, with $z_k = (\mu_{1,k}, \ldots, \mu_{Q,k}, \lambda_{1,k}, \ldots, \lambda_{k,k})$ and with $b_k = (0, \ldots, 0, f(x'_1), \ldots, f(x'_k))$.

Now adding a new point x'_{k+1} amounts to the following: First create a new matrix A_{k+1} obtained from A_k by adding a $(k + 1)$–th line and a $(k + 1)$-th column involving only some 0 and the $\varphi(\|x_i - x_{k+1}\|)$, $i = 1, \ldots, k + 1$. Second, solve the new linear system $A_{k+1} z_{k+1} = b_{k+1}$ for the new set z_{k+1} of coefficients, where b_{k+1} is just the former vector b_k for the first k components extended by another component $f_{k+1} = f(x'_{k+1})$.

Thus we can consider s_k as built from $X'_{k+1} = \{x'_1, x'_2, \ldots, x'_k, x'_{k+1}\}$, but with the coefficients $\lambda_{i,k+1} = \lambda_{i,k}$ for $i = 1, \ldots, k$ and $\lambda_{k+1,k+1} = 0$, and $\mu_{\ell,k+1} = \mu_{\ell,k}$ for $\ell = 1, \ldots, Q$, i.e., we refer to a vector $\tilde{z}_k = (z_k, 0)$.

Then, using that there exists a constant K depending on φ and Ω only such that $\|s_k - s_{k+1}\|_{\infty,\Omega} \leq K\|z_{k+1} - \tilde{z}_k\|_p$, it is easy to check whether the new point is significant or not.

4.1. The Bordering Method

Fortunately for us, there is a method for solving linear systems that is taylor–made for our purpose, namely the bordering method that we recall here for completeness [see 3,8]. This method is stongly connected with the Schur's complement of a square matrix, [18].

Let M be a matrix partitioned into four blocks,

$$M = \begin{pmatrix} A & B \\ C & D \end{pmatrix},$$

where the submatrix A is assumed to be square and regular. The Schur's complement of A with respect to M is defined as

$$(M/A) = D - CA^{-1}B.$$

In case M is a square matrix, this is directly connected with block Gaussian decomposition, since then it is obvious that

$$M = \begin{pmatrix} A & B \\ C & D \end{pmatrix} = \begin{pmatrix} I & 0 \\ CA^{-1} & I \end{pmatrix} \begin{pmatrix} A & B \\ 0 & (M/A) \end{pmatrix}.$$

Moreover, $\det M = \det A \times \det(M/A)$ [18].

Let us remark that having the nonsingular submatrix A in the upper left corner of M is not mandatory. We can as well define

$$(M/B) = C - DB^{-1}A, \quad (M/C) = B - AC^{-1}D, \quad (M/D) = A - BD^{-1}C,$$

assuming of course that B, C, or D, is a regular square submatrix of M. This provides us with the following decompositions:

$$M = \begin{pmatrix} I & 0 \\ DB^{-1} & I \end{pmatrix} \begin{pmatrix} A & B \\ (M/B) & 0 \end{pmatrix}, \quad \det M = -\det B \times \det(M/B).$$

$$M = \begin{pmatrix} I & AC^{-1} \\ 0 & I \end{pmatrix} \begin{pmatrix} 0 & (M/C) \\ C & D \end{pmatrix}, \quad \det M = -\det C \times \det(M/C).$$

$$M = \begin{pmatrix} I & BD^{-1} \\ 0 & I \end{pmatrix} \begin{pmatrix} (M/D) & 0 \\ C & D \end{pmatrix}, \quad \det M = \det D \times \det(M/D).$$

The bordering method for solving a linear system is easy to describe. Let A_k be a square regular matrix of dimension k, let a_k be a scalar, and let u_k and v_k be a column vector and a row vector, respectively, of dimension k. We consider the bordered matrix of dimension $k + 1$,

$$A_{k+1} = \begin{pmatrix} A_k & u_k \\ v_k & a_k \end{pmatrix}.$$

Then the Gaussian LU–factorization $A_k = L_k U_k$ can be updated to $A_{k+1} = L_{k+1} U_{k+1}$ by putting

$$\alpha_k = (A_{k+1}/A_k) = a_k - v_k A_k^{-1} u_k = \frac{\det A_{k+1}}{\det A_k},$$

and by using the following formulas:

$$L_{k+1} = \begin{pmatrix} L_k & 0 \\ v_k U_K^{-1} & 1 \end{pmatrix}, \quad U_{k+1} = \begin{pmatrix} U_k & L_k^{-1} u_k \\ 0 & \alpha_k \end{pmatrix},$$

$$L_{k+1}^{-1} = \begin{pmatrix} L_k^{-1} & 0 \\ -v_k A_k^{-1} & 1 \end{pmatrix}, \quad U_{k+1}^{-1} = \frac{1}{\alpha_k} \begin{pmatrix} U_k^{-1} & -A_k^{-1} u_k \\ 0 & 1 \end{pmatrix},$$

and

$$A_{k+1}^{-1} = \frac{1}{\alpha_k} \begin{pmatrix} \alpha_k A_k^{-1} + A_k^{-1} u_k v_k A_k^{-1} & -A_k^{-1} u_k \\ -v_k A_k^{-1} & 1 \end{pmatrix}.$$

Moreover, if z_k is the solution of $A_k z_k = b_k$, and if z_{k+1} satisfies $A_{k+1} z_{k+1} = b_{k+1}$, where $b_{k+1} = \begin{pmatrix} b_k \\ f_k \end{pmatrix}$ with $f_k \in \mathbb{R}$, it is obvious that we obtain the well–known formula [see 8]

$$z_{k+1} = \begin{pmatrix} z_k \\ 0 \end{pmatrix} = \frac{f_k - v_k \alpha_k}{\alpha_k} \begin{pmatrix} -A_k^{-1} u_k \\ 1 \end{pmatrix}.$$

For practical purposes, it is even possible to avoid the computation and the storage of the matrix $-A_k^{-1}$ since we may compute the vector $q_k = -A_k^{-1} u_k$ directly by a bordering method:

- Let k be fixed.

- For $i = 1, \ldots, k$, let $q_k^{(i)}$ denote the solution of $A_i q_k^{(i)} = -u_k^{(i)}$, where $u_k^{(i)}$ is the vector of the i first components of u_k, and $u_{k,i+1}$ is its $(k+1)$–th component (thus $u_i^{(i)} = u_i$ and $q_i^{(i)} = q_i$).

- By the bordering method, we can write:

$$\begin{cases} q_k^{(1)} = -\dfrac{u_k^{(1)}}{A_1}. \\[2ex] \begin{cases} \text{For } i = 1, \ldots, k-1: \\[1ex] q_k^{(i+1)} = \begin{pmatrix} q_k^{(i)} \\ 0 \end{pmatrix} - \dfrac{1}{a_i + v_i q_i^{(i)}} \left(u_{k,i+1} + v_i q_k^{(i)} \right) \begin{pmatrix} q_i^{(i)} \\ 1 \end{pmatrix}. \end{cases} \end{cases}$$

Then

$$q_k^{(k)} = q_k = -A_k^{-1} u_k .$$

Remark: In order to deal with more than a single point at each step, it is possible to consider also a block bordering method as follows: Let us look at the linear equation

$$A_{k+1} Z_{k+1} = F_{k+1}$$

where $F_{k+1} = \begin{pmatrix} F_k \\ \Phi_k \end{pmatrix}$ and $A_{k+1} = \begin{pmatrix} A_k & B_k \\ C_k & D_k \end{pmatrix}$, the size of the blocks being compatible. The Schur's complement (A_{k+1}/A_k) of A_k with respect to A_{k+1} is the matrix

$$(A_{k+1}/A_k) = \Lambda_k = D_k - C_k A_k^{-1} B_k ,$$

which we assume to be regular. Then

$$Z_{k+1} = \begin{pmatrix} Z_k \\ 0 \end{pmatrix} + \begin{pmatrix} A_k^{-1} B_k \\ -I \end{pmatrix} \Lambda_k^{-1} \left(C_k Z_k - \Phi_k \right) ,$$

with Z_k the solution of $A_k Z_k = F_k$. This amounts to the

Algorithm :

Let n denote the order of A_{k+1}.

- If $n < 4$, then solve the system directly.

- If $n \geq 4$, then do:

   ▷ Let $p = \lfloor \frac{n+1}{2} \rfloor$ and $q = n - p$.

   So, if A_k is $p \times p$, then D_k is $q \times q$.)

   ▷ Solve $A_k \begin{bmatrix} Z_k G_k \end{bmatrix} = \begin{bmatrix} F_k B_k \end{bmatrix}$.

   (So, $G_k = A_k^{-1} B_k$.)

   ▷ Let $E_k = C_k Z_k - \Phi_k$ and $\Lambda_k = D_k - C_k G_k$.

   ▷ Solve $\Lambda_k X_k = E_k$.

- Then $Z_{k+1} = \begin{pmatrix} Z_k + G_k X_k \\ -X_k \end{pmatrix}.$

References

[1]   A. Bouhamidi, A. Le Méhauté: *Multivariate interpolating* (m, l, s)-*splines,* Adv. Comput. Math. **11** (1999), 287–314.

[2]   A. Bouhamidi, A. Le Méhauté: *Spline curves and surfaces under tension,* in: Wavelets, Images and Surfaces Fitting, A. K. Peters, Wellesley 1994, pp. 51–58.

[3]   C. Brezinski: *Other manifestations of the Schur complement,* Université de Lille, ANO report 193 (1987).

[4]   M. D. Buhmann: *New developments in radial basis function interpolation,* in: Multivariate Approximation: From CAGD to Wavelets, K. Jetter, F. Utreras (eds.), World Scientific, Singapore 1993, pp. 35–75.

[5]   M. D. Buhmann, F. Derrien, A. Le Méhauté: *Spectral properties and knot removal for interpolation by pure radial sums,* in: Mathematics of Curves and Surfaces, M. Dæhlen, T. Lyche, L. L. Schumaker (eds.), Vanderbilt University Press, Nashville 1995, pp. 55–62.

[6]   M. D. Buhmann, A. Le Méhauté: *Knot removal with radial basis function interpolants,* Compt. Rend. Acad. Sci. Paris **320**, Sér. 1, vol. 4 (1995), 501–506.

[7]   J. Duchon: *Splines minimizing rotation–invariant semi–norms in Sobolev spaces,* in *Multivariate Approximation Theory,* W. Schempp, K. Zeller (eds.), Birkhäuser, Basel 1979, pp. 85–100.

[8]   D. K. Faddeev, V. N. Faddeeva: *Computational Methods of Linear Algebra,* W. H. Freeman, San Francisco 1963.

[9]   R. L. Hardy: *Multiquadric equations of topography and irregular surfaces,* J. Geophys. Res. **76** (1971), 1905–1915.

[10]   K. Jetter: *Conditionally lower Riesz bounds for scattered data interpolation,* in: Wavelets, Images and Surface Fitting, P.–J. Laurent, A. Le Méhauté, L. L. Schumaker (eds.), A. K. Peters, Wellesley 1994, pp. 295–302.

[11]   A. Le Méhauté: *Knot removal for Scattered Data,* in: Approximation Theory, Wavelets and Applications, S. P. Singh (ed.), NATO Advanced Study Inst. **454,** Kluwer, Dordrecht 1995, pp.197–213.

[12]   A. Le Méhauté, A. Bouhamidi: *Splines in approximation and dif-
        ferential operators: Interpolating (m, ℓ, s)-spline,* in: CRM Pro-
        ceedings and Lecture Notes **17**, S. Dubuc, (ed.), 1998, pp. 67–77.

[13]   A. Le Méhauté, Y. Lafranche: *A knot removal strategy for scat-
        tered data in $I\!R^2$,* in: Mathematical Methods in CAGD, T. Lyche,
        L. L. Schumaker (eds.), Academic Press, New York 1989, pp. 419–
        425.

[14]   T. Lyche, K. Mørken: *Knot removal for parametric B-splines
        curves and surfaces,* Comp. Aided Geom. Design **4** (1987), 217–
        230.

[15]   F. J. Narcowich, J. D. Ward: *Norm estimates for the inverses
        of a general class of scattered-data radial-function interpolation
        matrices,* J. Approx. Theory **69** (1992), 84–109.

[16]   R. Schaback: *Error estimates and condition numbers for radial
        basis function interpolation,* manuscript 1993.

[17]   R. Schaback: *Comparison of radial basis function interpolants,* in:
        Multivariate Approximation: From CAGD to Wavelets, K. Jetter,
        F. Utreras, eds, World Scientific, Singapore (1993), pp. 35–75.

[18]   I. Schur: *Über Potenzreihen, die im Innern des Einheitskreises
        beschränkt sind,* J. Reine Angew. Math. **147** (1917), 205–232.

Addresses:

Alain Le Méhauté
Département de Mathématiques
Faculté des Sciences
Université de Nantes
F–44072 Nantes
France

Yvon Lafranche
Laboratoire de Mathématiques Appliquées
Université de Rennes 1
F–35040 Rennes
France

International Series of Numerical Mathematics
Vol. 137, ©2001 Birkhäuser Verlag Basel/Switzerland

Tangent Space Methods for Approximation on Compact Homogeneous Manifolds

Jeremy Levesley

Abstract

Tangent space ideas are utilised to prove convergence of invariant kernel interpolation on compact homogeneous manifolds, as well as Taylor series type estimates for polynomial approximation on the same manifolds.

1 Introduction

In this paper I will try to persuade you that it is a good idea to use the tangent space when doing approximation on manifolds. The motivation for this came from attempts to produce high order error estimates for radial interpolation on the sphere. Efforts in this direction had stalled because it was not known how to produce scaling arguments for Lagrange interpolation on spheres. Jetter et. al. [5] solve this problem by using global data satisfying a point separation condition, and then using the Bernstein inequality. Arguments more tailored to local error estimates have now been made (see Bos and de Marchi [1] and v. Golitschek and Light [4]). We wanted to do something different and perhaps more natural, and that is to use tangent space ideas and the already well-known results concerning the scaling of Lagrange polynomials in Euclidean space.

We also became interested in the question of local polynomial approximation using spherical polynomials on spheres as a means to producing piecewise spline estimates on the sphere; see [7]. The method used to solve this question is easily generalisable to more abstract manifolds and forms the material of Section 3.

We shall mainly be considering interpolation on a compact homogeneous d–dimensional manifold M, which we shall realise in the following way. Fix a point $p \in \mathbb{R}^{d+r}$ and let $M = \{gp : g \in G\}$, where $G \leq SO(d+r)$ is a subgroup of the rotation group in $d+r$ dimensions. The important point here is that we have embedded the manifold in a high dimensional Euclidean space so that the group acting on the manifold is a group of linear maps. For example, we embed the usual torus $\Pi^2 = \Pi' \times \Pi'$ in $\mathbb{R}^4 = \mathbb{R}^2 \times \mathbb{R}^2$ rather than in $\mathbb{R}^3$. We also require that M be *reflexive* in the sense that, for each pair of points $x, y \in M$, there exists an element $g \in G$ such that $gx = y$ and $gy = x$. This is a stronger condition than the more usual transitivity condition which just requires there to be a $g \in G$ such that $gx = y$.

The manifold M comes equipped with a normalised G–invariant measure μ such that $\mu(M) = 1$.

We shall be using G–invariant kernels for interpolation. These are the natural analogue of radial kernels both in Euclidean space and on the sphere. We say that k is G–invariant if, for all $g \in G$, and $x, y \in M$, $k(gx, gy) = k(x, y)$. Suppose, for some continuous function f, we know function values $f(x_1), \ldots, f(x_N)$ at points $x_1, \ldots, x_N \in M$. Then, we look for an approximation to f of the form

$$s_k(x) \quad = \quad \sum_{i=1}^{N} \beta_i k(x, x_i), \tag{1.1}$$

where we determine the coefficients $\beta_1, \ldots, \beta_N$ (hopefully) using the interpolation conditions

$$s_k(x_j) \quad = \quad f(x_j), \qquad j = 1, \ldots, N. \tag{1.2}$$

The function s_k is called the k–spline interpolant to f.

As we shall refer to in passing later, if the matrix $[k(x_i, x_j)]_{i,j=1}^{N}$ resulting from the interpolation process is positive definite then we will have a unique solution to (1.2). We will put conditions on k to ensure that this happens.

In Section 4 we shall look at the rate of convergence of the k–spline interpolant to f as the interpolation points become dense on M, and give an error estimate that relates the smoothness of the kernel to the rate of convergence. We will start however with some necessary harmonic analysis on M, and follow this with the short Section 3 which deals with local polynomial approximation on M.

2   Harmonic Analysis on Manifolds

Let II_n be the space of polynomials of degree at most n in $d+r$ variables (recall that our manifold is homogeneously embedded in $\mathbb{R}^{d+r}$), and $P_n = \text{II}_n|_M$. Then, the *homogeneous polynomials* on M are $H_n = P_n \cap P_{n-1}^\perp$, where orthogonality is with respect to the G-invariant measure μ. It is straightforward to show that both P_n and H_n are G-invariant. We can uniquely decompose H_n into irreducible G-invariant subspaces H_{nk}, $k = 1, \ldots, \nu_n$. However, for the purposes of this paper it will be enough to assume that the space H_n is already irreducible, and so save ourselves a plethora of subscripts. Let $\{Y_{n,1}, \ldots, Y_{n,d_n}\}$ be an orthonormal basis for H_n.

Lemma 2.1. *Fix $0 \neq p_\lambda \in H_n$. Then,*

$$H_n = \text{span}\,\{p(g\cdot) : g \in G\}.$$

Proof: Since span $\{p(g\cdot) : g \in G\}$ is a non–trivial G–invariant subspace of H_n, which is irreducible, it must be the whole of H_n. □

We can develop a Fourier series expansion for any function $f \in L_1(M)$:

$$f \sim \sum_{n=0}^{\infty} \sum_{k=1}^{d_n} \alpha_{nk}(f) Y_{nk},$$

where

$$\alpha_{nk} := \int_M f Y_{nk}\, d\mu.$$

The Fourier series can thus be written

$$\begin{aligned}
f(y) &\sim \sum_{n=0}^{\infty} \sum_{k=1}^{d_n} \int_M f(x) Y_{nk}(x)\, d\mu(x) Y_{nk}(y) \\
&= \sum_{n=0}^{\infty} \int_M \left\{ \sum_{k=1}^{d_n} Y_{nk}(x) Y_{nk}(y) \right\} f(x)\, d\mu(x) \\
&= \sum_{n=0}^{\infty} \int_M p_n(x,y) f(x)\, d\mu(x),
\end{aligned}$$

where p_n is a *kernel of projection* onto H_n, and is thus a *reproducing kernel* for H_n.

Suppose k_1 and k_2 are kernels of projection onto H_n. Then $(k_1(\,,y), p) = (k_2(\,,y), p) = p$ for all $p \in H_n$. Thus $([k_1 - k_2](\,,y), p) = 0$ for all $p \in H_n$

and $y \in M$. Fix $y \in M$ and let $p = [k_1 - k_2](\,,y)$. Then, for each $y \in M$, $\| [k_1 - k_2](\,,y)\|_2^2 = 0$ so that $k_1 - k_2 = 0$. Thus we have

Corollary 2.2.   *For each $n = 0, 1, \ldots$, the kernel of projection onto H_n is unique. In particular, the kernel is independent of the original choice of orthonormal basis.*

Lemma 2.3. $p_n : M \times M \to \mathbb{R}$ *is a G–invariant kernel.*

Proof: Let $x, y \in M$. Acting on both variables of k simultaneously by an element of G we have

$$p_n(gx, gy) \;=\; \sum_{k=1}^{d_n} Y_{nk}(gx) Y_{nk}(gy)$$
$$=\; p_n(x, y),$$

by Lemma 2.2, since $\{Y_{nk}(g\cdot),\, k = 1, \ldots, d_n\}$ is an orthonormal basis for H_n.

$\square$

We now wish to decompose each G–invariant kernel into a sum of these fundamental G–invariant kernels. To this we first need to show:

Lemma 2.4.   *For a G–invariant kernel k let*

$$T_k f(y) = \int_M k(x, y) f(x) \, d\mu(x).$$

Then, if k_1 and k_2 are continuous G–invariant kernels, $T_{k_1} T_{k_2} f = T_{k_2} T_{k_1} f$ for all $f \in \mathbf{C}(M)$.

Proof: Let $f \in \mathbf{C}(M)$. Then, since the functions involved are all continuous and the manifold is compact we can use Fubini's theorem to reorder the integration

$$T_{k_1} T_{k_2} f(y) \;=\; \int_M k_1(x, y) \left\{ \int_M k_2(x, z) f(z) \, d\mu(z) \right\} d\mu(x)$$
$$=\; \int_M \int_M k_1(x, y) k_2(x, z) f(z) \, d\mu(x) \, d\mu(z)$$
$$=\; \int_M \left\{ \int_M k_1(x, y) k_2(x, z) \, d\mu(x) \right\} f(z) \, d\mu(z)$$
$$=\; \int_M \left\{ \int_M k_1(g^{-1}x, y) k_2(g^{-1}x, z) \, d\mu(x) \right\} f(z) \, d\mu(z),$$

where g here will depend on y and z, both of which may be fixed inside the integral. We have also used the G–invariance of μ. Using the G–invariance of the kernels, and then, using the reflexivity of M, choosing g so that $gz = y$ and $gy = z$, we have

$$
\begin{aligned}
T_{k_1} T_{k_2} f(y) &= \int_M \left\{ \int_M k_1(x, gy) k_2(x, gz) \, d\mu(x) \right\} f(z) \, d\mu(z) \\
&= \int_M \left\{ \int_M k_1(x, z) k_2(x, y) \, d\mu(x) \right\} f(z) \, d\mu(z) \\
&= \int_M k_2(x, y) \left\{ \int_M k_1(x, z) f(z) \, d\mu(z) \right\} d\mu(x) \\
&= T_{k_2} T_{k_1} f(y). \qquad \qquad \qquad \qquad \qquad \qquad \square
\end{aligned}
$$

Using the Lemma 2.4 we can show that the homogeneous polynomials are eigenfunctions of the integral operators associated with G–invariant kernels. This result is well known for the Laplace operator on the spheres.

Proposition 2.5. *Let k be a continuous G–invariant kernel. Then, for all $p \in H_n$*

$$
T_k p = \gamma_n p,
$$

where

$$
\gamma_n = \frac{1}{d_n} \int_M p_n(x, z) k(x, z) \, d\mu(z), \quad x \in M.
$$

(The integral is independent of x as we will see below.)

Proof: Since p_n is the reproducing kernel for H_n we have $T_{p_n} p = p$ for all $p \in H_n$. By Lemma 2.4 we see that, for all $p \in H_n$,

$$
\begin{aligned}
T_k p &= T_k T_{p_n} p \\
&= T_{p_n} T_k p \in H_n,
\end{aligned}
$$

since T_{p_n} projects onto H_n. Thus, T_k is a linear operator from H_n to itself. The G–invariant kernel k is symmetric since $k(x, y) = k(gy, gx) = k(y, x)$ for some $g \in G$, using the reflexivity of M. Thus, T_k restricted to H_n is self–adjoint. Therefore p. Let $\pi \in H_n$ be arbitrary. Then, by Lemma 2.1, $\pi(x) = \sum_{i=1}^{d_n} \beta_i p(g_i x)$ for some $g_i \in G$, and $\beta_i \in \mathbb{R}$, $i = 1, \ldots, d_n$. Then,

$$
\begin{aligned}
T_k \pi(x) &= \int_M k(x, y) \left\{ \sum_{i=1}^{d_n} \beta_i p(g_i y) \right\} d\mu(y) \\
&= \sum_{i=1}^{d_n} \beta_i \int_M k(x, y) p(g_i y) \, d\mu(y)
\end{aligned}
$$

$$= \sum_{i=1}^{d_n} \beta_i \int_M k(x, g_i^{-1}y)p(y)\, d\mu(y)$$

$$= \sum_{i=1}^{d_n} \beta_i \int_M k(g_i x, y)p(y)\, d\mu(y)$$

$$= \gamma_n \sum_{i=1}^{d_n} \beta_i p(g_i x)$$

$$= \gamma_n \pi(x).$$

We have used the G–invariance both of μ and k in the argument above. Thus, all elements of H_n are eigenvectors of T_k, with eigenvalue γ_n.

For the final part of the proposition fix $x \in M$. Then, $p_n(x, \cdot) \in H_n$ so that

$$\int_M p_n(x, y)k(x, y)\, d\mu(y) = \gamma_n p_n(x, x).$$

Since p_n is G–invariant $p_n(x, x)$ is independent of x. Now,

$$p_n(x, x) = \int_M p_n(x, x)\, d\mu(x)$$

$$= \int_M \left\{ \sum_{k=1}^{d_n} Y_{nk}(x)Y_{nk}(x) \right\} d\mu(x) = d_n,$$

since the Y_{nk} are orthonormal. Substituting this into the previous equation gives

$$\gamma_n = \frac{1}{d_n} \int_M p_n(x, x)\, d\mu(x). \qquad \square$$

Theorem 2.6. *Let k be a continuous G–invariant kernel. Then, in $L_2(M)$,*

$$k(x, y) = \sum_{n=0}^{\infty} \gamma_n p_n(x, y),$$

where γ_n is given in Proposition 2.5.

Proof: If we fix $x \in M$, $k(x, \cdot)$ is a univariate function and has a Fourier series expansion

$$k(x, y) = \sum_{n=0}^{\infty} \int_M p_n(y, z)k(x, z)\, d\mu(z)$$

$$= \sum_{n=0}^{\infty} \gamma_n p_n(x, y),$$

using Proposition 2.5 and the symmetry of p_n. $\qquad \square$

3 Local Polynomial Approximation

In this section we prove a local polynomial approximation result. The basis of the proof is the multivariate Taylor series estimate for $f \in C^k(B_{\rho_0}(y))$, where, for $y \in \mathbb{R}^d$, $B_\rho(y) = \{x : \|x - y\| \leq \rho\}$. For then there is a polynomial $p_{k-1} \in \Pi_{k-1}$ such that, for every $\rho \leq \rho_0$,

$$\|p_{k-1} - f\|_{\infty, B_\rho(y)} \leq C\|f^{(k)}\|_{\infty, B_{\rho_0}(y)}\rho^k. \tag{3.1}$$

The manifold M is d–dimensional, but is embedded in $\mathbb{R}^{d+r}$. Let $\{e_1, \ldots, e_d\}$ be an orthonormal basis for the tangent space at $x \in M$ (x is the origin of the tangent space). Extend this to an orthonormal basis $\{e_1, \ldots, e_{d+r}\}$ for the ambient Euclidean space $\mathbb{R}^{d+r}$. Then, since M is a manifold there is, for $\rho \leq \rho_0$ with ρ_0 sufficiently small, and each point $z = z_1 e_1 + \ldots + z_d e_d \in B_\rho(0)$ a unique point $y \in M$ with first d coordinates $z_1, \ldots, z_d$. This is the orthogonal projection $\pi : B_\rho(0) \to M$. For $x, y \in M$ let $d(x, y)$ be the distance on M between x and y. Let $D_\rho(x) = \{y : d(y, x) \leq \rho\}$. Since Euclidean space is flat $D_\rho(x) \subseteq \pi(B_\rho(0))$.

So, suppose we begin with a function $f \in C^k(D_\eta(x))$. Then, $f \circ \pi \in \mathbf{C}^k(B_\rho(0))$. Using (3.1) we know there is a polynomial $p_{k-1} \in \Pi_{k-1}$ such that

$$\|f \circ \pi - p_{k-1}\|_{\infty, B_\rho(0)} \leq C\|f \circ \pi\|_{\infty, B_{\rho_0}(0)}\rho^k$$
$$\leq C\|f \circ \pi\|_{\infty, B_{\rho_0}(0)}\eta^k.$$

Though p_{k-1} is a polynomial in d variables, we can view it as a polynomial in $d + r$ variables just by setting all the other variables to 0. In this context (with a slight abuse of notation), since π is the orthogonal projection onto M, $p_{k-1} \circ \pi^{-1} = p_{k-1}|_M$. Then, since $D_\rho(x) \subseteq \pi(B_\rho(0))$,

$$\|f - p_{k-1} \circ \pi^{-1}\|_{\infty, D_\rho(x)} = \|f \circ \pi - p_{k-1}\|_{\infty, B_\rho(0)}$$
$$\leq C\|(f \circ \pi)^{(k)}\|_{\infty, B_{\rho_0}(0)}\rho^k.$$

Thus we have the following local polynomial approximation theorem:

Theorem 3.1. *Let $f \in \mathbf{C}^k(D_{\rho_0}(x))$ for some $\rho_0 > 0$. Then, there exists a polynomial $p_{k-1} \in P_{k-1}$ such that, for every $\rho \leq \rho_0$,*

$$\|f - p_{k-1}\|_{\infty, D_\rho(x)} \leq C\rho^k,$$

where the constant C depends on f but is independent of ρ.

4   Convergence Rates for Invariant Kernel Approximation

The convergence rates we give in this section are all based on the well–known reproducing kernel Hilbert space method of e.g. Duchon [2], Golomb and Weinberger [3], and Wahba [8]. We give a brief review of the main ideas here, the goal being to arrive at the *power function* estimate Theorem 4.3 below.

We begin with a G–invariant continuous kernel, which, by virtue of Theorem 2.6, has the expansion

$$k(x,y) \;=\; \sum_{n=0}^{\infty} a_n p_n(x,y),$$

where all of the coefficients $a_n > 0$. Such a kernel is positive definite in the sense that

$$\sum_{i,j=1}^{N} c_i c_j k(x_i, x_j) \;>\; 0,$$

for all $N \in \mathbb{N}$, distinct $x_1, \ldots, x_N \in M$, and $c_1, \ldots, c_N \in \mathbb{R}$. This ensures that the interpolation equations (1.2) are uniquely solvable, and also allows for the following definition of an inner product. Let

$$(f,g)_k \;=\; \sum_{n=0}^{\infty} a_n^{-1} \sum_{k=1}^{d_n} \alpha_{nk}(f) \alpha_{nk}(g).$$

We are now in a position to define the space of functions which we can approximate:

$$X_k = \{ f \in L_2(M) : \|f\|_k := (f,f)_k^{1/2} < \infty \}.$$

Then, if we impose the condition

$$\sum_{n=0}^{\infty} d_n a_n < \infty,$$

we have $X_k \subset \mathbf{C}(M)$ and interpolation makes sense.

Lemma 4.1.   *The kernel k is the reproducing kernel for X_k, i.e., for all $f \in X_k$ and $x \in M$,*

$$(f, k(x, \cdot))_k = f(x).$$

Proof: Because

$$k(x, y) = \sum_{n=0}^{\infty} a_n p_n(x, y) \quad = \quad \sum_{n=0}^{\infty} a_n \sum_{j=1}^{d_n} Y_{nj}(x) Y_{nj}(y),$$

$$(f, k(x, \cdot))_k \quad = \quad \sum_{n=0}^{\infty} a_n^{-1} \sum_{j=1}^{d_n} (a_n Y_{nj}(x)) \alpha_{nj}(f)$$

$$= \quad \sum_{n=0}^{\infty} \sum_{j=1}^{d_n} \alpha_{nj}(f) Y_{nj}(x)$$

$$= \quad f(x). \qquad \square$$

A direct consequence of this lemma is the following orthogonality result.

Corollary 4.2. *Let s_k be the k–spline interpolant to $f \in X_k$ at the points $x_1, \ldots, x_N \in M$. Then*

(i) $(f - s_k, s_k)_k = 0$,

(ii) $\|f\|_k^2 = \|f - s_k\|_k^2 + \|s_k\|^2$,

(iii) s_k *is the interpolant to f from X_k of minimum norm.*

Proof:

(i) Since

$$s_k(x) = \sum_{i=1}^{N} \beta_i k(x, x_i)$$

for some $\beta_1, \ldots, \beta_N \in \mathbb{R}$,

$$(f - s_k, s_k)_k \quad = \quad \sum_{i=1}^{N} \beta_i (f - s_k, k(\cdot, x_i))_k$$

$$= \quad \sum_{i=1}^{N} \beta_i (f(x_i) - s_k(x_i))$$

$$= \quad 0, \qquad (4.1)$$

since s_k interpolates f at $x_1, \ldots, x_N$.

(ii) This follows directly from (i) using Pythagoras' theorem.

(iii) As in part (i) we see that, if $g(x_i) = f(x_i)$, $i = 1, \ldots, N$, for $g \in X_k$, Equation (4.1) holds with $f = g$. Thus, $\|s\|_k \le \|g\|_k$, so that s_k is the interpolant from X_k of minimum norm. $\qquad \square$

We may use these results to prove the usual error estimate for interpolation.

Theorem 4.3. *Let s_k interpolate $f \in X_k$ at $x_1, \ldots, x_N$. Then, for each $x \in M$,*

$$|f(x) - s_k(x)|$$
$$\le \min_{c_1, \ldots, c_N \in \mathbb{R}} \|f\|_k \left\{ k(x, x) - 2 \sum_{i=1}^{N} c_i k(x, x_i) + \sum_{i,j=1}^{N} c_i c_j k(x_i, x_j) \right\}^{1/2}.$$

Proof: Using the reproducing kernel property of k we have

$$\begin{aligned} |f(x) - s_k(x)| &= (f - s_k, k(x, \cdot))_k \\ &= (f - s_k, k(x, \cdot) + \sum_{i=1}^{N} c_i k(x_i, \cdot))_k \\ &\le \|f - s_k\| \| k(x, \cdot) + \sum_{i=1}^{N} c_i k(x_i, \cdot)\|_k \\ &\le \|f\| \| k(x, \cdot) + \sum_{i=1}^{N} c_i k(x_i, \cdot)\|_k, \end{aligned}$$

where we have used the reproducing kernel property Lemma 4.1, the Cauchy–Schwarz inequality, and Lemma 4.2 (ii). Note that the coefficients $c_1, \ldots, c_N$ are arbitrary. We now use the fact that $\| \cdot \|_k = (\cdot, \cdot)_k^{1/2}$ and the reproducing kernel property of k to rewrite the last inequality

$$|f(x) - s_k(x)|$$
$$\le \|f\|_k \left\{ k(x, x) - 2 \sum_{i=1}^{N} c_i k(x, x_i) + \sum_{i,j=1}^{N} c_i c_j k(x_i, x_j) \right\}^{1/2}.$$

The result follows when we minimise the last expression over all possible choices of $c_1, \ldots, c_N$. $\qquad \square$

The bracketed expression in the last equation above is often referred to as the *power function*.

In order to use this error estimate we need to make an effective choice of coefficients $c_1, \ldots, c_N$. In order to do this we need to recast the problem on the tangent space $\mathbb{R}^d$ at x. For convenience we will denote by I the interpolation set $\{x_1, \ldots, x_N\}$. We will assume that we have a compact set $\Gamma \subset \mathbb{R}^d$ and an infinitely differentiable chart $\psi : \Gamma \to V := \psi(\Gamma) \subset M$ with $x \in V$. Let $I_\Gamma = I \cap V$, and let $Z = \{z_1, \ldots, z_L\} = \psi^{-1}(I_\Gamma)$. Then Z is the set of preimages under ψ of the interpolation points that lie in $\psi(\Gamma)$.

Since ψ is a differentiable function with differentiable inverse there exist constants C_1, C_2 such that

$$C_1\|z - w\| \leq d(\psi(z), \psi(w)) \leq C_2\|z - w\|, \qquad (4.2)$$

so that distances on M and in $\mathbb{R}^d$ are boundedly equivalent.

Before we give the next result we need to provide some notation. Let $M_k = \dim(\Pi_k)$ (here we will assume that we are working with polynomials in d dimensions). Then, a set of points $\{\xi_1, \ldots, \xi_{M_k}\}$ is Π_k–unisolvent if we can construct a unique Lagrange polynomials basis , i.e. we have a set of polynomials $\{p_1, \ldots, p_{M_k}\}$ which satisfy $p_i(\xi_j) = \delta_{ij}, 1 \leq i, j, \leq M_k$.

Proof of the following crucial result may be found in Madych and Nelson [6].

Proposition 4.4. *Let U be a bounded subset of $\mathbb{R}^d$ and $\Xi = \{\xi_1, \ldots, \xi_L\} \subset U$ satisfying*

$$\rho = \max_{z \in V} \min_{\xi \in \Xi} \|z - \xi_i\|.$$

Then, there exists a subset $U_\rho \subset U$ with $\operatorname{diam}(U_\rho) \leq A\rho$, where A is independent of ρ, so that we can select a Π_k–unisolvent subset $\Xi_\rho \subset \Xi \cup U_\rho$. Furthermore, if $p_1^\rho, \ldots, p_{M_k}^\rho$ are the Lagrange polynomials for Ξ_ρ the selection can be made so that

$$\max_{z \in U} \max_{i=1,\ldots,M_k} |p_i^\rho(z)| \leq D,$$

where the constant D is independent of ρ.

We can now prove the main result of this section.

Theorem 4.5. *Let k be a G–invariant kernel which is $2r$ times continuously differentiable in each variable. Let s_k be the k–spline interpolant to a function $f \subset X_k$ at the point set $I = \{x_1, \ldots, x_N\}$, which satisfies*

$$\max_{x \in M} \min_{y \in I} d(x, y) = \eta.$$

Then, there exists $\eta_0 > 0$ so that whenever $\eta \leq \eta_0$

$$\|f(x) - s_k(x)\|_{\infty,M} \leq C\|f\|_k\eta^r.$$

Proof: Fix $x \in V \subset M$, where ψ, Γ, V and Z are as described above. Then, using (4.2), we see that Z is a set such that

$$\max_{w \in \Gamma} \min_{z \in Z} \|w - z\| \leq \rho := \eta/C_1.$$

Using Proposition 4.4 we can find a $U_\rho \subset \Gamma$, with $\mathrm{diam}\,(U_\rho) \leq A\rho$ (A independent of ρ), $Z_\rho \subset Z \cup U_\rho$ (which we number $z_1, \ldots, z_{M_{2r-1}}$) such that Z_ρ is Π_{2r-1}–unisolvent. If $p_1^\rho, \ldots, p_{M_{2r-1}}^\rho$ are the Lagrange polynomials for $z_1, \ldots, z_{M_{2r-1}}$,

$$\max_{w \in U_\rho} \max_{i=1,\ldots,M_{2r-1}} |p_i^\rho(w)| \leq D, \tag{4.3}$$

where the constant D is independent of ρ.

We now choose the coefficients

$$c_i = \begin{cases} -p_i^\rho(0), & i = 1, \ldots, M_{2r-1}, \\ 0, & \text{otherwise.} \end{cases}$$

For then we have, for any $q \in \Pi_{2r-1}$,

$$\sum_{i=1}^{N} c_i q(x_i) = -q(0). \tag{4.4}$$

The kernel $\kappa(z, w) = k(\psi(z), \psi(w))$ is $2r$–times continuously differentiable in each variable. Viewing (z, w) as a single $2d$–dimensional variable we can do a Taylor series approximation to κ:

$$\kappa(z, w) = t_{2r-1}(z, w) + R_{2r}(z, w), \tag{4.5}$$

where the Taylor series remainder

$$\begin{aligned} |R_{2r}(z, w)| &\leq C_k\|(z, w)\|^{2r} \\ &\leq C_k(\|z\| + \|w\|)^{2r}, \end{aligned} \tag{4.6}$$

and C_k depends on the kernel k, but is independent of z and w, since these are confined to a compact set. We note that $t_{2r-1}(0,0) = \kappa(0,0)$

Using (4.5) we can now write the error estimate of Theorem 4.3 in the form

$$|f(x) - s_k(x)|$$

$$\leq \|f\|_k \left\{ \kappa(0,0) - 2 \sum_{i=1}^{M_{2r}} c_i \kappa(0, z_i) + \sum_{i,j=1}^{M_{2r}} c_i c_j k(z_i, z_j) \right\}^{1/2}$$

$$= \|f\|_k \left\{ \kappa(0,0) - 2 \sum_{i=1}^{M_{2r}} c_i(t_{2r-1}(0, z_i) + R_{2r}(0, z_i)) \right.$$

$$\left. + \sum_{i,j=1}^{M_{2r}} c_i c_j(t_{2r-1}(z_i, z_j) + R_{2r}(z_i, z_j)) \right\}^{1/2}$$

$$= \|f\|_k \left\{ \kappa(0,0) - 2\kappa(0,0) - 2 \sum_{i=1}^{M_{2r}} c_i(R_{2r}(0, z_i)) \right.$$

$$\left. - \sum_{i=1}^{2r} c_i t_{2r-1}(z_i, 0) + \sum_{i,j=1}^{M_{2r}} c_i c_j R_{2r}(z_i, z_j)) \right\}^{1/2}$$

$$= \|f\|_k \left\{ -\kappa(0,0) + \kappa(0,0) - 2 \sum_{i=1}^{M_{2r}} c_i(R_{2r}(0, z_i)) + \sum_{i,j=1}^{M_{2r}} c_i c_j R_{2r}(z_i, z_j)) \right\}^{1/2},$$

where we have used (4.4) in the last two lines above. If we now use the bound on the remainder (4.6), and the bound $\mathrm{diam}\,(U_\rho) \leq A\rho$, we see that

$$|f(x) - s_k(x)| \leq \|f\|_k \left\{ D M_{2r-1} C_k (A\rho)^{2r} + 2^{2r} D^2 M_{2r-1} C_k (A\rho)^{2r} \right\}^{1/2}$$
$$\leq C\|f\|_k \eta^r,$$

where C depends on the set U but not on η. $\qquad\square$

References

[1]   L. Bos, S. de Marchi: *Limiting values under scaling of the Lebesgue function for polynomial interpolation on spheres*, J. Approx. Theory **96** (1999), 366–377.

[2]   J. Duchon: *Sur l'erreur d'interpolation des fonctions de plusieurs variables par les D^m-splines*, RAIRO Anal. Num. **12** (1978), 325–334.

[3]   M. Golomb, H. F. Weinberger: *Optimal approximation and error bounds,* in: R. E. Langer (Ed): On Numerical Approximation, Univ. Wisconsin Press, Madison 1959, pp. 117–190.

[4]   M. von Golitschek, W. A. Light: *Interpolation by polynomials and radial basis functions on spheres,* Constr. Approx. **17** (2001), 1–18.

[5]   K. Jetter, J. Stöckler, J. D. Ward: *Error estimates for scattered data interpolation on spheres,* Math. Comp. **68** (1999), 733–747.

[6]   W. R. Madych, S. A. Nelson: *Multivariate interpolation and conditionally positive definite functions II,* Math. Comp. **54** (1990), 211–230.

[7]   L. L. Schumaker, T. Lyche: *A multiresolution tensor spline method for fitting functions on the sphere,* SIAM J. Sci. Comput. **22** (2000), 724–746.

[8]   G. Wahba: *Spline interpolation and smoothing on the sphere,* SIAM J. Sci. Statist. Comput. **2** (1981), 5–16, and **3** (1982), 385–386.

Address:

Jeremy Levesley
Department of Mathematics and Computer Science
University of Leicester
Leicester LE1 7RH
England

International Series of Numerical Mathematics
Vol. 137, ©2001 Birkhäuser Verlag Basel/Switzerland

Open Problem Concerning Fourier Transforms of Radial Functions in Euclidean Space and on Spheres

Jeremy Levesley and Simon Hubbert

Let $P_n^{(\lambda)}$ be the Gegenbauer polynomials, orthogonal on $[-1,1]$ with respect to the weight $(1-t^2)^{\lambda-1/2}$, normalised by

$$P_n^{(\lambda)}(1) = \frac{\Gamma(n+2\lambda)}{\Gamma(2\lambda)\Gamma(n+1)}.$$

A function is said to be completely monotone on $(0,\infty)$ if $(-1)^k \phi^{(k)}(t) > 0$ for all $t \in (0,\infty)$.

Let ϕ be continuous on $[0,\infty)$, and $\phi^{(m)}$ be completely monotone on $(0,\infty)$ for some $m \in \mathbb{N}$. For $x \in \mathbb{R}^d$, let $\Phi(x) = \phi(\|x\|)$ have a (generalised) Fourier transform with polynomial decay, that is

$$\hat{\Phi}(\xi) = \mathcal{O}(\|\xi\|^{-d-\alpha}),$$

for some $\alpha > 0$ and $d = 2\lambda + 2$.

Conjecture. *The restriction of Φ to the sphere,*

$$\Psi(x,y) = \phi\left(\sqrt{2 - 2x^T y}\right), \qquad x,y \in S^{d-1},$$

has a representation as a spherical Fourier series

$$\Psi(x,y) = \sum_{k=0}^{\infty} \sum_{l=1}^{d_k} c_k Y_{kl}(x) Y_{kl}(y),$$

whose spherical Fourier coefficients $\{c_k\}$ have the analagous decay rate

$$c_k = \mathcal{O}(k^{-d+1-\alpha}).$$

Here $\{Y_{kl}\}$, $l = 1,\ldots,d_k$, $k = 0,1,\ldots$, form an orthonormal basis for the spherical harmonics of degree k, which has dimension d_k.

References

[1]   B. J. C. Baxter, S. Hubbert: *Radial basis functions for the sphere.* In: Recent Progress in Multivariate Approximation, Internat. Ser. Numer. Math. **137**, W. Haußmann, K. Jetter, M. Reimer (eds.), Birkhäuser, Basel 2001, pp. 33–47.

[2]   E. W. Cheney, W. A. Light: *A Course in Approximation Theory*, Pacific Grove, California, 2000.

[3]   E. M. Stein, G. Weiss: *Introduction to Fourier Analysis on Euclidean Spaces*, Princeton University Press, Princeton, N. J., 1990.

[4]   J. Levesley: *Convergence of Euclidean radial basis approximation on spheres*, Technical Report 2000/28, Dept. of Mathematics and Computer Science, University of Leicester, Leicester LE1 7RH, 2000.

[5]   J. Levesley, W. A. Light, D. L. Ragozin, X. Sun: *A simple approach to the variational theory for interpolation on spheres.* In: New Developments in Approximation Theory, Interat. Ser. Numer. Math. **132**, M. W. Müller, M. D. Buhmann, D. H. Mache, M. Felten (eds.), Birkhäuser, Basel 1999, pp. 117–143.

Addresses:

Jeremy Levesley
Department of Mathematics and Computer Science
University of Leicester
Leicester LE1 7RH
England

Simon Hubbert
Department of Mathematics
Imperial College
London SW7 2BZ
England

International Series of Numerical Mathematics
Vol. 137, © 2001 Birkhäuser Verlag Basel/Switzerland

Local Lagrange Interpolation on Powell–Sabin Triangulations and Terrain Modelling

Günther Nürnberger and Frank Zeilfelder

Abstract

Local Lagrange interpolation schemes for quadratic C^1−splines on arbitrary triangulations with Powell–Sabin splits are constructed. By using the concept of weak interpolation, it is proved that the interpolation method yields optimal approximation order. We test our method by interpolating scattered data and show how the method can be applied for terrain modelling. We compare the interpolating splines on fine and coarse triangulations obtained from thinning strategies and analyze the data reduction.

Introduction

There is a vast literature on local Hermite interpolation respectively quasi–interpolation (cf. Argyis, Fried and Scharpf [1], Chui and Hong [2], Clough and Tocher [3], Dahmen, Gmelig Meyling and Ursem [4], Davydov, Nürnberger and Zeilfelder [8], Heindl [12], Laghchim–Lahlou and Sablonnière [15, 16], Lai and Schumaker [17, 18, 19], Morgan and Scott [20], Powell and Sabin [33], Sablonnière [34], Ursem [35]) and on non–local Lagrange interpolation (cf. Davydov and Nürnberger [5], Davydov, Nürnberger and Zeilfelder [6, 7], Nürnberger and Rießinger [22, 23], Nürnberger and Walz [24], Nürnberger and Zeilfelder [25, 26]). On the other hand, only recently, the first local Lagrange interpolation methods for C^1–splines of degree at least three on triangulations were developed (Nürnberger and Zeilfelder [28, 29], Nürnberger, Schumaker and Zeilfelder [30, 31]). For further references we refer to the survey [27].

In this paper we contruct local Lagrange interpolation schemes for quadratic C^1–splines on arbitrary triangulations, where each triangle is subdivided by a Powell–Sabin split. We emphasize that the classical Hermite interpolation method of Powell and Sabin [33] cannot by transformed directly into a local Lagrange interpolation scheme. Therefore, we have to describe an algorithm how to choose the Lagrange interpolation points on the edges of the triangulation. Since its complexity is linear in the number of triangles and the fundamental splines have compact support, the interpolating spline can be computed very fast. By applying the concept of weak interpolation, it is proved that the interpolation method yields optimal approximation order.

Local Lagrange interpolation methods are important for the construction and reconstruction of surfaces, since only data is needed and no derivatives. For example, if in practice, a surface is described by a linear spline on a fine triangulation (with many triangles), then the interpolating C^1–spline on a coarse subtriangulation can be constructed by taking the Lagrange data directly from the linear spline. In our computational tests, the coarse piecewise linear splines are obtained by applying the thinning strategies of Dyn, Floater and Iske [11] developed recently. We test our method by interpolating geological scattered data which show that the method works efficiently for terrain modelling. We compare the interpolating C^1–splines on fine and coarse triangulations and discuss the results in the context of data reduction.

1   Preliminaries and Notations

Let a regular triangulation Δ of a simply connected polygonal domain $\Omega \subset \mathbb{R}^2$ be given (i.e., a set of closed triangles such that the intersection of any two triangles is empty, a common edge or a vertex). We consider the so–called *Powell–Sabin refinement* Δ_{PS} of Δ. This triangulation Δ_{PS} is obtained from Δ as follows: first for each triangle $T \in \Delta$ the center v_T of its inscribed circle is connected to the three vertices of T. Then the centers of inscribed circles of neighboring triangles of Δ are connected to each other, and, finally, for each triangle $T \in \Delta$ that has a boundary edge e, the center of its inscribed circle is connected to the midpoint of e (see Figure 1.). We note that this construction leads to a *singular vertex* of Δ_{PS} on each interior edge of Δ. (As usual, an interior vertex of Δ_{PS} is called singular, if there exist only two edges with different slopes that are attached to it.) The existence of such an intersection point is guaranteed since the splitting point v_T is always chosen as the center of the inscribed circle, but other choices for the splitting point may also be appropriate.

Throughout this paper, for each edge $e = [u, v]$ of Δ we denote by $v_e = v_{[u,v]}$ the vertex of Δ_{PS} lying in the interior of e (*i.e.* $v_e = v_{[u,v]}$ is a vertex of Δ_{PS} which is not a vertex of Δ) and we set α for the smallest angle of the triangles in Δ. Moreover, we denote by h the maximal diameter of the triangles in Δ_{PS}.

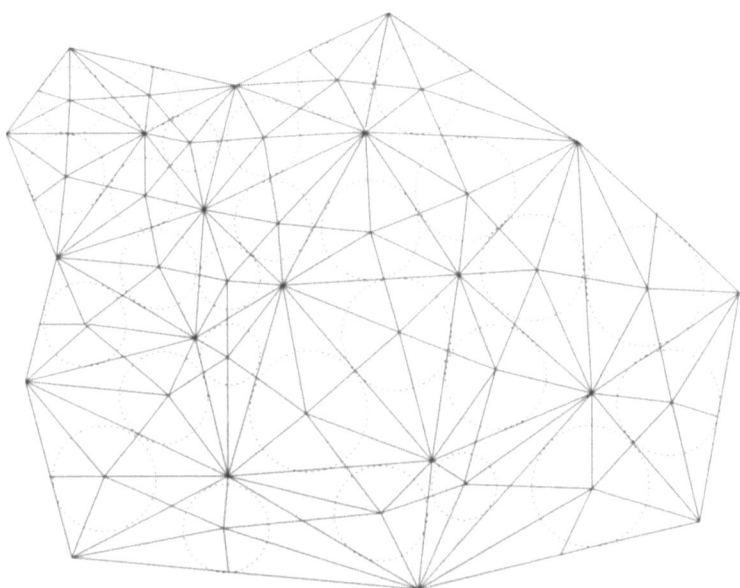

Figure 1: The Powell–Sabin refinement Δ_{PS} of a triangulation Δ.

We consider the space of *quadratic C^1–splines* with respect to Δ_{PS}. These spaces are defined by

$$S_2^1(\Delta_{PS}) = \{s \in C^1(\Omega) : \ s|_T \in \Pi_2, \ T \in \Delta_{PS}\},$$

where

$$\Pi_2 = \ span\{x^i y^j : i, j \geq 0, \ i + j \leq 2\}$$

is the 6 dimensional space of *quadratic polynomials*. It is well–known that the dimension of the space $S_2^1(\Delta_{PS})$ equals $3V$, where V is the number of vertices of Δ.

In the next section, we give an algorithm to construct local Lagrange interpolation sets for C^1–splines for $S_2^1(\Delta_{PS})$.

A subset $\{z_1, \ldots, z_{3V}\}$ of Ω is called *Lagrange interpolation set* for $S_2^1(\Delta_{PS})$, if for each function $f \in C(\Omega)$, a unique spline $s_f \in S_2^1(\Delta_{PS})$ exists that satifies the *Lagrange interpolation conditions*

$$s_f(z_\nu) = f(z_\nu), \ \nu = 1, \ldots, 3V.$$

A Lagrange interpolation set for $S_2^1(\Delta_{PS})$ is called a *local Lagrange interpolation set* for $S_2^1(\Delta_{PS})$, if the corresponding Lagrange fundamental splines $s_\mu \in S_2^1(\Delta_{PS})$, $\mu = 1, \ldots, d$, defined by

$$s_\mu(z_\nu) = \delta_{\mu,\nu}, \ \nu = 1, \ldots, 3V,$$

have local support ($\delta_{\mu,\nu}$ denotes Kronecker's symbol). This means that the support of s_μ,

$$supp(s_\mu) = \overline{\{z \in \Omega : \ s_\mu(z) \neq 0\}},$$

is contained in some m–star of a triangle $T \in \Delta$. Here, we define the closed star of a triangle T, denoted by $\mathrm{Star}_\Delta^1(T) \subseteq \Omega$, as the union of all triangles in Δ having a common vertex with T. And we define the m–star of T, denoted by $\mathrm{Star}_\Delta^m(T) \subseteq \Omega$, $m \geq 2$, as the union of all triangles in Δ that intersect with $\mathrm{Star}_\Delta^{m-1}(T)$ (see Figure 2.).

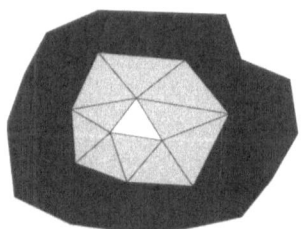

Figure 2: Definition of the m–star of T. Adding the grey triangles to the white triangle T gives $\mathrm{Star}_\Delta^1(T)$. Adding the dark triangles to $\mathrm{Star}_\Delta^1(T)$ gives $\mathrm{Star}_\Delta^2(T)$.

In addition, we use the following notations. Given a sufficiently differentiable function f, we denote by $D_r f$ the partial derivatives of f in direction of the unit vector r. For $r = (1,0)$ and $r = (0,1)$, respectively, we write $D_x f$ and $D_y f$, respectively. Given r_1 and r_2, two unit vectors, the higher partial derivatives are denoted by $D_{r_1^i r_2^j} f$. In this case, the number $\omega = i + j$ is called the degree of $D_{r_1^i r_2^j} f$. Given a point $z = (x, y) \in \Omega$, we set

$$D^\omega f(z) = (D_{x^\omega} f(z), D_{x^{\omega-1}y} f(z), \ldots, D_{xy^{\omega-1}} f(z), D_{y^\omega} f(z)).$$

The uniform norm of f is defined by $\|f\| = \max\{|f(z)| : \ z \in \Omega\}$ and for the derivatives, we set

$$\|D^\omega f\| = \max\{\|D_{x^i y^j} f\| : \ i \geq 0, \ j \geq 0, \ i + j = \omega\}.$$

2 Local Lagrange Interpolation

In this section, we describe our algorithm for constructing local Lagrange interpolation sets for $S_2^1(\Delta_{PS})$. We state and prove our main results on local Lagrange interpolation by $S_2^1(\Delta_{PS})$ and the approximation properties of the interpolating spline.

We start by describing the algorithm. The construction of local Lagrange interpolation sets $\mathcal{L}$ for $S_2^1(\Delta_{PS})$ is based on adding successively Lagrange interpolation points in the interior of the edges of Δ to $\mathcal{V}$, the set of vertices of Δ.

Construction of the interpolation set $\mathcal{L}$. Let $T_1, \ldots, T_n$ be the triangles of Δ. During the algorithm we successively mark the vertices of Δ (whenever interpolation points are chosen near the vertices).

First, we choose the points

$$u, \ v, \ w, \ \tfrac{1}{2}(v_{[u,v]} + u), \ \tfrac{1}{2}(v_{[u,w]} + u), \ \tfrac{1}{2}(v_{[v,w]} + v),$$
$$\tfrac{1}{2}(v_{[u,v]} + v), \ \tfrac{1}{2}(v_{[u,w]} + w), \ \tfrac{1}{2}(v_{[v,w]} + w),$$

where u, v, w are the vertices of T_1, and mark the vertices u, v, w. Then, we proceed by induction as follows. Let us assume that we have considered the triangles $T_1, \ldots, T_i$. Now, we consider the triangle T_{i+1}, and again denote its vertices by u, v, w. We consider two cases. If the vertices u, v, w are unmarked, then we choose the points

$$u, \ v, \ w, \ \tfrac{1}{2}(v_{[u,v]} + u), \ \tfrac{1}{2}(v_{[u,w]} + u), \ \tfrac{1}{2}(v_{[v,w]} + v),$$
$$\tfrac{1}{2}(v_{[u,v]} + v), \ \tfrac{1}{2}(v_{[u,w]} + w), \ \tfrac{1}{2}(v_{[v,w]} + w).$$

Otherwise, we omit T_{i+1}, and consider T_{i+2}.

If we have considered $T_1, \ldots, T_n$, then we denote the unmarked vertices by $u_1, \ldots, u_m$. We first consider u_1, and distinguish between two cases. If an unmarked vertex u_j, $j \geq 2$, exists such that $[u_1, u_j]$ is an edge of Δ, then we choose an arbitrary triangle T with vertices u_1, u_j, u. In this case, we choose the interpolation points

$$u_1, \ u_j, \ \tfrac{1}{2}(v_{[u_1,u_j]} + u_1), \ \tfrac{1}{2}(v_{[u_1,u]} + u_1), \ \tfrac{1}{2}(v_{[u_j,u_1]} + u_j), \ \tfrac{1}{2}(v_{[u_j,u]} + u_j),$$

and mark the vertices u_1, u_j. Otherwise, we choose an arbitrary triangle T with vertices u_1, u, v. In this case, we choose the interpolation points

$$u_1, \ \tfrac{1}{2}(v_{[u_1,u]} + u_1), \ \tfrac{1}{2}(v_{[u_1,v]} + u_1),$$

and mark the vertex u_1. Then, we proceed by induction as follows. Let us assume that we have considered the vertices $u_1, \ldots, u_i$. Now, we consider the vertex u_{i+1}. If u_{i+1} is marked, then we omit u_{i+1}, and consider u_{i+2}. Otherwise, if an unmarked vertex u_j, $j \geq i+2$, exists such that $[u_{i+1}, u_j]$ is an edge of Δ, then we choose an arbitrary triangle T with vertices u_{i+1}, u_j, u. In this case, we choose the interpolation points

$$u_{i+1}, \; u_j, \; \tfrac{1}{2}(v_{[u_{i+1},u_j]}+u_{i+1}), \; \tfrac{1}{2}(v_{[u_{i+1},u]}+u_{i+1}), \; \tfrac{1}{2}(v_{[u_j,u_{i+1}]}+u_j), \; \tfrac{1}{2}(v_{[u_j,u]}+u_j),$$

and mark the vertices u_{i+1}, u_j. If such a vertex u_j does not exist, then we choose an arbitrary triangle T with vertices u_{i+1}, u, v. In this case, we choose the interpolation points

$$u_{i+1}, \; \tfrac{1}{2}(v_{[u_{i+1},u]} + u_{i+1}), \; \tfrac{1}{2}(v_{[u_{i+1},v]} + u_{i+1}),$$

and mark the vertex u_{i+1}.

The set of chosen interpolation points is denoted by $\mathcal{L}$.

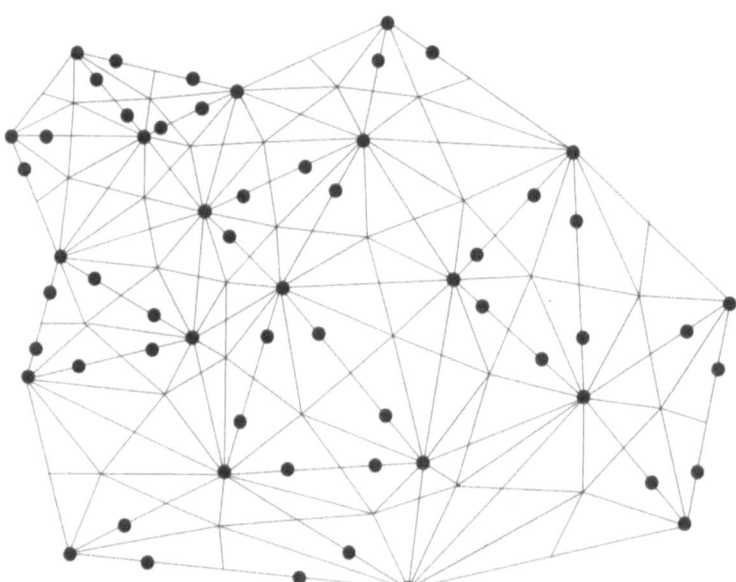

Figure 3: A local Lagrange interpolation set for $S_2^1(\Delta_{PS})$: the Lagrange interpolation points are marked by filled circles.

The above algorithm leads to a Lagrange interpolation set $\mathcal{L}$ for $S_2^1(\Delta_{PS})$. We note that Lagrange type interpolation for certain subspaces of $S_2^1(\Delta_{PS}^1)$, where Δ^1 is the uniform three directional mesh in the (whole) plane, was investigated in [14]. The above Lagrange interpolation set $\mathcal{L}$ consists of $3V$ Lagrange interpolation points. An example for $\mathcal{L}$ is given in Figure 3. We note that the algorithmic complexity of this construction is linear in the number of triangles of Δ, which can be seen easily by Euler's formulas.

Theorem 2.1. *The set $\mathcal{L}$ is a local Lagrange interpolation set for $S_2^1(\Delta_{PS})$.*

Proof of Theorem 2.1: Let $\mathcal{L} \subset \Omega$ be the set of points constructed by the above algorithm, and let $s \in S_2^1(\Delta_{PS})$ be a spline with

$$s(z) = 0, \ z \in \mathcal{L}.$$

We have to show that $s \equiv 0$.

First, it is well–known from the univariate theory that the set

$$\{a, \tfrac{1}{2}(a + x_1), \tfrac{1}{2}(b + x_1), b\}$$

is a Lagrange interpolation set for the space of quadratic C^1–splines on $[a, b]$ with one interior knot $x_1 \in (a, b)$. Therefore, it follows that $s|_e \equiv 0$ holds for all edges e of Δ containing four Lagrange interpolation points.

We next consider all edges that contain exactly three points of $\mathcal{L}$. Let $e = [u, v]$ be such an edge and assume that $u, \tfrac{1}{2}(v_{[u,v]} + v), \ v \in \mathcal{L}$. By our method two adjacent edges e_1, e_2 of Δ with endpoint u exist that contain at least three points from $\mathcal{L}$. If the edges e_1, e_2 both contain four points of $\mathcal{L}$ then it follows that the gradient of s vanishes at u. In this case, it follows that $s|_e \equiv 0$. Otherwise, if (at least) one of the edges e_1, e_2 contains only three points from $\mathcal{L}$, then we obtain from the above argument that s vanishes on this edge. Again the gradient at u vanishes, and we obtain $s|_e \equiv 0$. A similar argument shows that $s|_e \equiv 0$ also holds for all edges that contain exactly two points of $\mathcal{L}$.

Now, it follows from our method that we have $s|_e \equiv 0$ for all edges e of Δ. Therefore, we obtain that the functional value and the gradient of s vanish for all vertices $v \in \mathcal{V}$. The classical result of Powell and Sabin [33] now implies $s \equiv 0$. Since, we have only used local arguments the proof is complete. □

In the following, we show that the local Lagrange interpolating spline constructed by the above algorithm yields optimal approximation order. For doing this, we use the next lemma on weak interpolation that follows from a general result in [29].

Here, we let $z_1, \ldots, z_6$, be (not necessarily different) points from a triangle $T \in \Delta_{PS}$ and we assign integers i_k, $j_k \geq 0$, $k = 1, \ldots, 6$. Moreover, if $i_k > 0$ and $j_k > 0$, then $r_{1,k}$ and $r_{2,k}$, respectively, are unit vectors in direction of an edge of T, which are required to be linearly independent, if $i_k > 0$ and $j_k > 0$. Moreover, we assume that for each sufficiently differentiable function $g : T \mapsto \mathbb{R}$ the polynomial interpolation problem

$$D_{r_{1,k}^{i_k} r_{2,k}^{j_k}} p(z_k) = D_{r_{1,k}^{i_k} r_{2,k}^{j_k}} g(z_k), \quad k = 1, \ldots, 6, \tag{2.1}$$

has a unique solution $p \in \Pi_2$.

Lemma 2.2. *Let $T \in \Delta_{PS}$ and $f \in C^3(T)$. If $p \in \Pi_2$ satisfies*

$$|D_{r_{1,k}^{i_k} r_{2,k}^{j_k}} (f - p)(z_k)| \leq \tilde{C} h^{3 - i_k - j_k}, \quad k = 1, \ldots, 6, \tag{2.2}$$

where $\tilde{C} > 0$ is a constant depending only on f and α, then there exists $C > 0$ (depending only on f and α) such that for all $\omega \in \{0, 1, 2\}$,

$$\|D^\omega (f - p)|_T\| \leq C h^{3 - \omega}.$$

In addition, we need the following auxiliary lemma on univariate splines.

Lemma 2.3. *Let $f \in C^3[a, b]$ and $S_2^1(\{x_1\})$ be the space of quadratic C^1-splines on $[a, b]$ with one interior knot $x_1 \in (a, b)$. Suppose that $s \in S_2^1(\{x_1\})$ satisfies one of the following properties:*

$$(i) \quad (f - s)(x) = 0, \ x \in \{a, \tfrac{1}{2}(a + x_1), \tfrac{1}{2}(b + x_1), b\}$$

$$(ii) \quad |(f - s)^{(j)}(x)| \leq \tilde{c} \delta^{3 - j}, \ j = 0, 1, \ x \in \{a, b\},$$

where $\tilde{c} > 0$ and $\delta = \max\{x_1 - a, b - x_1\}$. Then, there exists a constant $c > 0$ such that

$$|(f - s)(x_1)| \leq c \delta^3$$

Proof of Lemma 2.3: Let us first assume that (i) holds. It is well–known from univariate spline theory (cf. [21]), that there exist $\tilde{s} \in S_2^1(\{x_1\})$ such that

$$\|f - \tilde{s}\| \leq c_0 \delta^3.$$

Let $l_{i,[a,b]} \in S_2^1(\{x_1\})$, $i = 1, \ldots, 4$, be the Lagrange fundamental splines w.r.t. $\{t_1, t_2, t_3, t_4\} = \{a, \tfrac{1}{2}(a + x_1), \tfrac{1}{2}(b + x_1), b\}$, i.e. $l_{i,[a,b]}(t_j) = \delta_{i,j}$, $j = 1, \ldots, 4$. We have

$$|(f - s)(x_1)| \leq |(f - \tilde{s})(x_1)| + |(\tilde{s} - s)(x_1)| \leq c_0 \delta^3 + \sum_{i=1}^{4} |(\tilde{s} - s)(t_i)| \|l_{i,[a,b]}(x_1)|.$$

The desired inequation now follows from $\|l_{i,[a,b]}\| = \|l_{i,[0,1]}\|$ and (i), since $(\tilde{s} - s)(t_i) = (\tilde{s} - f)(t_i)$.

If (ii) holds, then we argue differently. Given a funtion $g \in C^3[\beta, \gamma]$, it follows from the Taylor expansion of g and g' that

$$2g(\beta) + g'(\beta)(\gamma - \beta) = 2g(\gamma) - g'(\gamma)(\gamma - \beta) + \tfrac{3g'''(\xi_1) - 2g'''(\xi_2)}{6}(\gamma - \beta)^3, \quad (2.3)$$

where $\xi_k \in (\beta, \gamma)$, $k = 1, 2$. Therefore, we obtain from (ii) a system in the unknowns $(f - s)(x_1)$, $(f - s)'(x_1)$ of the following type:

$$2(f - s)(x_1) - (f - s)'(x_1)\lambda\delta = \mathcal{O}(\delta^3)$$
$$2(f - s)(x_1) + (f - s)'(x_1)\tau\delta = \mathcal{O}(\delta^3),$$

where $\lambda\delta = x_1 - a$, $\tau\delta = b - x_1$. Cramer's rule now yields

$$|(f - s)(x_1)| = \tfrac{\delta\mathcal{O}(\delta^3)}{2(\lambda+\tau)\delta} = \mathcal{O}(\delta^3) .$$

This proves the lemma. □

Theorem 2.4 shows that the local Lagrange interpolating spline from $S_2^1(\Delta_{PS})$ and its derivatives have optimal approximation properties. In this theorem, the norm denotes the maximum of the uniform norm over all triangles of Δ_{PS} (w.r.t. the polynomial pieces).

Theorem 2.4. *Let $f \in C^3(\Omega)$ and $s_f \in S_2^1(\Delta_{PS})$, be the unique local Lagrange interpolating spline with $s_f(z) = f(z)$, $z \in \mathcal{L}$. Then, there exists a constant $K > 0$ (depending only on f and α), such that for all $\omega \in \{0, 1, 2\}$,*

$$\|D^\omega(f - s_f)\| \leq Kh^{3-\omega}. \quad (2.4)$$

Proof of Theorem 2.4: Let $s_f \in S_2^1(\Delta_{PS})$ be the unique Lagrange interpolation spline of f. Let $e_0 = [u, v]$ be an arbitrary edge of a triangle $T \in \Delta$ and denote by r_0 a unit vector in direction of e_0. Moreover, we set r_1 to be a unit vector in direction of the edge $[v_e, v_T]$ in Δ_{PS} and consider the univariate polynomials $p_1 = s_f|_{[u,v_e]}$, $p_2 = s_f|_{[v,v_e]}$. It follows from the Taylor expansion of the functions $D_{r_1}(f - p_1)|_{[u,v_e]}$, $D_{r_1}(f - p_2)|_{[v,v_e]}$ that

$$D_{r_1}(f - p_1)(u) = D_{r_1}(f - p_1)(v_e) - D_{r_1r_0}(f - p_1)(v_e)\lambda_1 h + \tfrac{D_{r_1r_0^2}f(\xi_1)(\lambda_1 h)^2}{2},$$

$$D_{r_1}(f - p_2)(v) = D_{r_1}(f - p_2)(v_e) + D_{r_1r_0}(f - p_2)(v_e)\lambda_2 h + \tfrac{D_{r_1r_0^2}f(\xi_2)(\lambda_2 h)^2}{2},$$

where $\xi_1 \in (u, v_e)$ and $\xi_2 \in (v, v_e)$ and $\lambda_i \in (0,1)]$, $i = 1,2$. Since v_e is a singular vertex of Δ_{PS}, we have

$$D_{r_1 r_0}(f - p_1)(v_e) = D_{r_1 r_0}(f - p_2)(v_e) .$$

Therefore, we get

$$
\begin{aligned}
D_{r_1}(f - s_f)(v_e) &= \tfrac{\lambda_1}{\lambda_1+\lambda_2}(D_{r_1}(f - s_f)(v) + \tfrac{\lambda_2}{\lambda_1}D_{r_1}(f - s_f)(u) \\
&\quad - \tfrac{\lambda_1\lambda_2^2 D_{r_1 r_0^2} f(\xi_2)+\lambda_1^2\lambda_2 D_{r_1 r_0^2} f(\xi_1)}{2\lambda_1}h^2) .
\end{aligned} \tag{2.5}
$$

From Lemma 2.2, Lemma 2.3, and the arguments given in the proof of Theorem 2.1, it follows that there exists a constant $\tilde{C}_0 > 0$ (depending only on f and α), such that for all edges e of Δ and for all $\omega \in \{0,1,2\}$,

$$\|D_{r^\omega}(f - s_f)|_e\| \le \tilde{C}_0 h^{3-\omega}, \tag{2.6}$$

where r is a unit vector in direction of e. Therefore, it follows from (2.5) that there exists a constant $\tilde{C}_1 > 0$ (depending only on f and α), such that for all edges e,

$$|D_{r_1}(f - s_f)(v_e)| \le \tilde{C}_1 h^2, \tag{2.7}$$

where r_1 is a unit vector as above.

We now show that there exists a constant $\tilde{C}_2 > 0$ (depending only on f and the smallest angle in Δ_{PS}) such that for every $T \in \Delta$,

$$|(f - s_f)(v_T)| \le \tilde{C}_2 h^3. \tag{2.8}$$

Let $T \in \Delta$ be a triangle with vertices u, v, and w. First, we consider the following sets

$$\mathcal{P}_1 = \{v_{[u,v]}, v_T, v_{[v,w]}\}, \quad \mathcal{P}_2 = \{v_{[u,v]}, v_T, v_{[u,w]}\}, \quad \mathcal{P}_3 = \{v_{[u,w]}, v_T, v_{[v,w]}\} .$$

If there exists $i \in \{1,2,3\}$ such that all the points in $\mathcal{P}_i$ lie on one single line, then (2.8) follows from (2.6), (2.7) and Lemma 2.3 (with assumption (ii)). Otherwise, we denote by r_1, r_2, r_3, a unit vector in direction of the edge $e_1 = [v_T, v_{[u,v]}]$, $e_2 = [v_T, v]$, $e_3 = [v_T, v_{[v,w]}]$, respectively, and we set θ_1, θ_2, for the angle between r_1 and r_2, r_2 and r_3, respectively. It follows from the relation (2.3), (2.6), and (2.7) that

$$2(f - s_f)(v_T) + D_{r_i}(f - s_f)(v_T)\tau_i h = \mathcal{O}(h^3), \quad i = 1,2,3,$$

where $\tau_i \in (0,1]$, $i = 1, 2, 3$. Moreover, we have

$$D_{r_3}(f - s_f)(v_T) = \frac{\sin(\theta_1 + \theta_2)}{\sin \theta_1} D_{r_2}(f - s_f)(v_T) - \frac{\sin \theta_2}{\sin \theta_1} D_{r_1}(f - s_f)(v_T).$$

These equations yield a linear system $Ax = b$, with $x^t = (x_1, x_2, x_3)$, with unknowns $x_1 = (f - s_f)(v_T)$, $x_2 = D_{r_1}(f - s_f)(v_T)$, and $x_3 = D_{r_2}(f - s_f)(v_T)$, where

$$A = \begin{pmatrix} 2 & \tau_1 h & 0 \\ 2 & 0 & \tau_2 h \\ 2 & -\frac{\sin \theta_2}{\sin \theta_1} \tau_3 h & \frac{\sin(\theta_1 + \theta_2)}{\sin \theta_1} \tau_3 h \end{pmatrix}.$$

We note that A is non–singular, since for $\theta_1 + \theta_2 \neq \pi$,

$$\frac{2h^2}{\sin \theta_1} (\sin \theta_1 \tau_1 \tau_2 + \sin \theta_2 \tau_2 \tau_3 - \sin(\theta_1 + \theta_2) \tau_1 \tau_3) \neq 0.$$

The desired inequation (2.8) is now obtained by applying Cramer's rule.

In [19], Lemma 2.2., it was shown that the smallest angle in Δ_{PS} can be bounded below by a constant depending α. Therefore, it follows from (2.6), (2.7), and (2.8) that for every triangle $T \in \Delta_{PS}$ with vertices u, v_e, and v_r there exists a constant $\tilde{C} > 0$ (depending only on f and α) such that

$$\max\{|(f - s_f)(u)|, |(f - s_f)(v_e)|, |(f - s_f)(v_r)|\} \leq \tilde{C} h^3$$

and

$$\max\{|D_x(f - s_f)(u)|, |D_y(f - s_f)(u)|, |D_{r_1}(f - s_f)(v_e)|\} \leq \tilde{C} h^2.$$

Since we have only used local arguments, Lemma 2.2. implies inequality (2.4) for all triangles $T \in \Delta_{PS}$ with the same constant K independent of h. The proof is complete. $\qquad \square$

3 Application to Terrain Modelling

In this section, we apply our method to interpolation of scattered geological data. The corresponding terrain surface is described as a piecewise linear spline on a fine triangulation with many triangles. We consider coarse versions of this piecewise linear surface which we obtain by applying the *thinning* strategies of Dyn, Floater and Iske [11]. In order to test our method, we interpolate these piecewise linear splines by quadratic C^1–splines. The C^1–splines are defined on a coarse subtriangulation which we get from constructing a constrained Delaunay triangulation of randomized points. The necessary Lagrange data is taken directly from the piecewise linear spline obtained from thinning. In this context, we discuss aspects of *data reduction*.

Figure 4 shows the original set of scattered data points from a Norwegian region. The terrain is described by a piecewise linear interpolant of the data. We consider the piecewise linear surfaces that are obtained by the adaptive thinning algorithm of Dyn, Floater and Iske [11] as applied to the data of Figure 4. Figure 5 shows the resulting piecewise linear surface which interpolates 7697 significant points of the original data set.

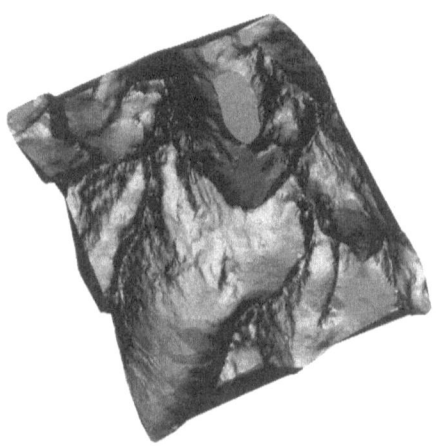

Figure 4: The original data consisting of 23092 points.

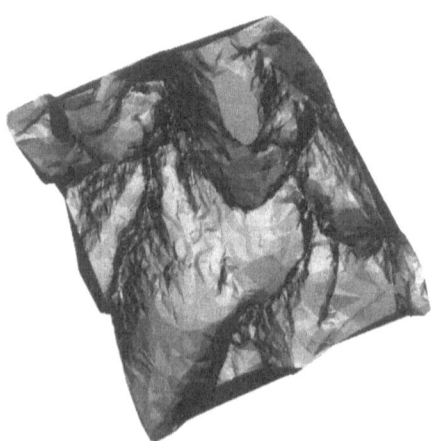

Figure 5: The piecewise linear spline interpolating 7697 data points.

We now interpolate the piecewise linear surface of Figure 5 by a quadratic C^1–spline. The corresponding Lagrange interpolation set consists of 7401 points. The resulting smooth surface is shown in Figure 6. As a further test, we interpolate the surface of Figure 5 by a quadratic C^1–spline at 3813 points (see Figure 7). In this case the compression rate is about 6.

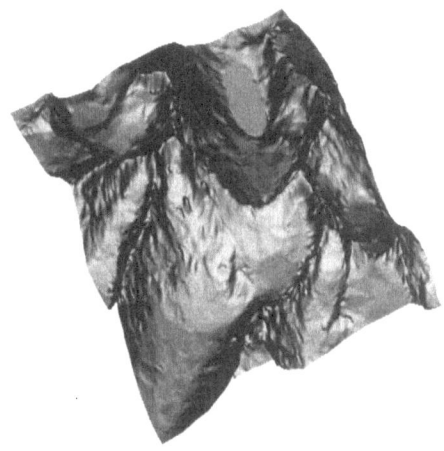

Figure 6: Lagrange Interpolation by quadratic C^1–splines at 7401 points.

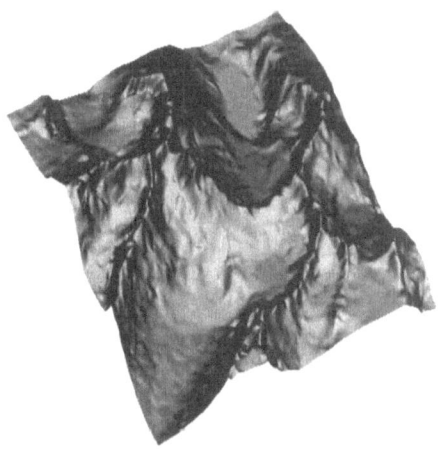

Figure 7: Lagrange Interpolation by quadratic C^1–splines at 3813 points.

We now consider the piecewise linear surface obtained by thinning which interpolates 2566 significant points of the original data set (see Figure 8). Figure 9 shows the quadratic C^1–spline which interpolates this surface at 2325 points. In this case, the compression rate is about 10.

Figure 8: The piecewise linear spline interpolating 2566 data points.

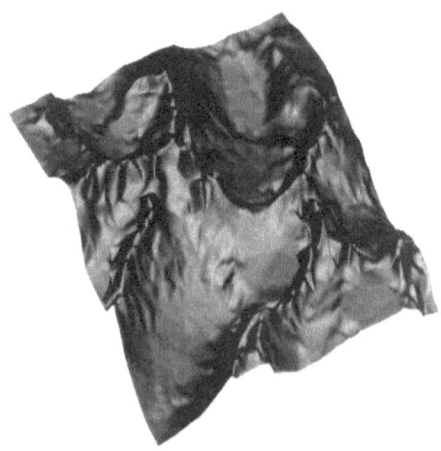

Figure 9: Lagrange Interpolation by quadratic C^1–splines at 2325 points.

Acknowledgment. The authors would like to thank Nira Dyn, University of Tel Aviv, for her idea of applying thinning in the context of smooth spline Lagrange interpolation, and Armin Iske, Technical University of Munich, for making his scattered data thinning software available to us.

References

[1]   J. H. Argyis, I. Fried, D. W. Scharpf: *The TUBA family of plate elements for the matrix displacement method,* Aeronaut. J. Roy. Aeronaut. Soc. **72** (1968), 701–709.

[2]   C. K. Chui, D. Hong: *Swapping edges of arbitrary triangulations to achieve the optimal order of approximation,* SIAM J. Numer. Anal. **34** (1997), 1472–1482.

[3]   R. W. Clough, J. L. Tocher: *Finite element stiffness matries for analysis of plates in bending,* in: Proc. Conf. on Matrix Methods in Structural Mechanics, Wright Patterson A. F. B., Ohio, 1965.

[4]   W. Dahmen, R. H. J. Gmelig Meyling, J. H. M. Ursem: *Scattered data interpolation by bivariate C^1-piecewise quadratic functions,* Approx. Theory Appl. **6** (1990), 6–29.

[5]   O. Davydov, G. Nürnberger: *Interpolation by C^1-splines of degree $q \geq 4$ on triangulations,* J. Comput. Appl. Math. **126** (2000), 159–183.

[6]   O. Davydov, G. Nürnberger, F. Zeilfelder: *Approximation order of bivariate spline interpolation for arbitrary smoothness,* J. Comp. Appl. Math. **90** (1998), 117–134.

[7]   O. Davydov, G. Nürnberger, F. Zeilfelder: *Cubic spline interpolation on nested polygon triangulations,* in: Curve and Surface Fitting, A. Cohen, C. Rabut, L. L. Schumaker (eds.), Vanderbilt University Press, Nashville, 2000, 161–170.

[8]   O. Davydov, G. Nürnberger, F. Zeilfelder: *Bivariate spline interpolation with optimal approximation order,* Constr. Approx. **17** (2001), 181–208.

[9]   N. Dyn, D. Levin, S. Rippa: *Data dependent triangulations for piecewise linear interpolation,* IMA J. Numer. Anal. **10** (1990), 137–154.

[10]   N. Dyn, T. Lyche: *A Hermite subdivision scheme for the eval-
         uation of the Powell–Sabin 12–split element*, in: Approximation
         Theory IX, C. K. Chui, L. L. Schumaker (eds.), Vanderbilt Uni-
         versity Press, Nashville, TN, 1998, 1–6.

[11]   N. Dyn, M. S. Floater, A. Iske: *Adaptive thinning for bivariate
         scattered data*, Technical University of Munich (2000), preprint.

[12]   G. Heindl: *Interpolation and approximation by piecewise quadratic
         C^1–functions of two variables*, in: Multivariate Approximation
         Theory, W. Schempp, K. Zeller (eds.), Birkhäuser, Basel, 1979,
         146–161.

[13]   P. Lancaster, K. Salkauskas: *Curve and Surface Fitting – an In-
         troduction*, Academic Press, London, 1986.

[14]   M. Laghchim–Lahlou: *The C^r–fundamental splines of Clough–
         Tocher and Powell–Sabin types for Lagrange interpolation on a
         three directional mesh*, Adv Comp. Math. **8** (1998) 353–366.

[15]   M. Laghchim–Lahlou, P. Sablonnière, *Triangular finite elements
         of HCT type and class C^ρ*, Adv. in Comp. Math. **2** (1994),
         101–122.

[16]   M. Laghchim–Lahlou, P. Sablonnière: *C^r–finite elements of
         Powell–Sabin type on the three directional mesh*, Adv. Comp.
         Math **6** (1996), 191–206.

[17]   M.–J. Lai, L. L. Schumaker: *On the Approximation Power of
         Bivariate Splines*, Adv. in Comp. Math. **9** (1998), 251–279.

[18]   M.–J. Lai, L. L. Schumaker: *Macro–elements and stable local bases
         for splines on Clough–Tocher triangulations*, (2000), preprint.

[19]   M.–J. Lai, L. L. Schumaker: *Macro–elements and stable local bases
         for splines on Powell–Sabin triangulations*, (2000), preprint.

[20]   J. Morgan, R. Scott: *A nodal basis for C^1 piecewise polynomials
         of degree $n \geq 5$*, Math. Comp. **29** (1975) 736–740.

[21]   G. Nürnberger: *Approximation by Spline Functions*, Springer,
         Berlin, 1989.

[22]   G. Nürnberger, T. Rießinger: *Lagrange and Hermite interpolation by bivariate splines*, Numer. Funct. Anal. Optim. **13** (1992) 75–96.

[23]   G. Nürnberger, T. Rießinger: *Bivariate spline interpolation at grid points*, Numer. Math. **71** (1995) 91–119.

[24]   G. Nürnberger, G. Walz: *Error analysis in interpolation by bivariate C^1-splines*, IMA J. Numer. Anal. **18** (1998) 485–508.

[25]   G. Nürnberger, F. Zeilfelder, *Interpolation by spline spaces on classes of triangulations*, J. Comput. Appl. Math. **119** (2000), 347–376.

[26]   G. Nürnberger, F. Zeilfelder, *Lagrange interpolation by splines on triangulations*, in: Proceedings of the Morningside Institute, R. H. Wang (ed.), Peking, 1998.

[27]   G. Nürnberger, F. Zeilfelder: *Developments in bivariate spline interpolation*, J. Comput. Appl. Math. **121** (2000), 125–152.

[28]   G. Nürnberger, F. Zeilfelder: *Local Lagrange interpolation by cubic splines on a class of triangulations*, Proc. Conf. Trends in Approximation Theory, Nashville 2000, in press.

[29]   G. Nürnberger, F. Zeilfelder: *Lagrange interpolation by bivariate C^1-splines with optimal approximation order*, in preparation.

[30]   G. Nürnberger, L. L. Schumaker, F. Zeilfelder: *Local Lagrange interpolation by bivariate C^1 cubic splines*, Proc. Conf. Curves and Surfaces, Oslo 2000, in press.

[31]   G. Nürnberger, L. L. Schumaker, F. Zeilfelder: *Local Lagrange interpolation by bivariate C^1-splines on convex quadrangulations*, in preparation.

[32]   G. Nürnberger, O. Davydov, G. Walz, F. Zeilfelder: *Interpolation by bivariate splines on crosscut partitions*, in: Multivariate Approximation and Splines, G. Nürnberger, J. W. Schmidt, G. Walz (eds.), Internat. Ser. Numer. Math. **125**, Birkhäuser, Basel, 1997, 189–204.

[33]   M. J. D. Powell, M. A. Sabin: *Piecewise quadratic approximation on triangles*, ACM Trans. Math. Software **4** (1977), 316–325.

[34]   P. Sablonnière: *Error bounds for Hermite interpolation by quadratic splines on an α−Triangulation,* IMA J. Numer. Anal. **7** (1987), 495–508.

[35]   J. H. M. Ursem: *Interpolation of bivariate functions by quadratic C^1-splines,* Doctoral thesis, University of Amsterdam, 1986.

Address:

Günther Nürnberger, Frank Zeilfelder
Fakultät für Mathematik und Informatik
Universität Mannheim
D–68131 Mannheim
Germany

International Series of Numerical Mathematics
Vol. 137, ©2001 Birkhäuser Verlag Basel/Switzerland

The Geometry of Nodes in a Positive Quadrature on the Sphere

Manfred Reimer

Assume the integral I,

$$IF := \int_{S^{r-1}} F(x)d\omega(x), \quad F \in C(S^{r-1}), \tag{1}$$

ω standard measure on the unit sphere S^{r-1} in $\mathbb{R}^r$, $r \geq 3$, is approximated by the quadrature $\hat{I}$,

$$\hat{I}F := \sum_{j=1}^{M} A_j F(t_j), \quad F \in C(S^{r-1}), \tag{2}$$

with nodes $t_1, ..., t_M \in S^{r-1}$ and weights $A_j > 0$ and satisfying

$$\hat{I}F = IF \quad \text{for all} \quad F \in \mathbb{P}_\mu^r, \tag{3}$$

$\mathbb{P}_\mu^r$ the space of all spherical polynomials of degree μ, $\mu \in \mathbb{N}$. What can be said about the geometry of the nodes on the basis of this information?

To show that some statements are possible without any additional assumptions we give several examples. In their formulation we need the following definitions: Let $T := \{t_1, ..., t_M\}$ and let

$$P_\mu := \frac{1}{\omega_{r-1}}(C_\mu^{\frac{r}{2}} + C_{\mu-1}^{\frac{r}{2}}) = \text{const} \cdot P_\mu^{(\frac{r-1}{2}, \frac{r-3}{2})}, \tag{4}$$

$$G_\mu := \frac{2\mu + r - 2}{(r-2)\omega_{r-1}} \cdot C_\mu^{\frac{r-2}{2}} \tag{5}$$

be the reproducing kernel functions of $\mathbb{P}_\mu^r$ and of its subspace $\mathbb{H}_\mu^r$ consisting of all spherical harmonics of degree μ, respectively. This means that

$$F(x) = \int_{S^{r-1}} F(t)P_\mu(xt)d\omega(t) \quad \text{for all} \quad F \in \mathbb{P}_\mu^r, x \in S^{r-1},$$

$$H(x) = \int_{S^{r-1}} H(t)G_\mu(xt)d\omega(t) \quad \text{for all} \quad H \in \mathbb{H}_\mu^r, x \in S^{r-1}$$

hold. The zeros $\eta_{\mu k} = \cos \chi_{\mu k}$ of G_μ and $\bar{\eta}_{\mu k} = \cos \bar{\chi}_{\mu k}$ of P_μ are ordered by

$$0 < \chi_{\mu 1} < \dots < \chi_{\mu \mu} < \pi, \quad 0 < \bar{\chi}_{\mu 1} < \dots < \bar{\chi}_{\mu \mu} < \pi.$$

And we introduce the spherical collars

$$C(t; \phi_0, \phi_1) := \{x \in S^{r-1} \mid \cos \phi_1 \leq tx \leq \cos \phi_0\} \tag{6}$$

for $t \in S^{r-1}$, $0 \leq \phi_0 < \phi_1 \leq \pi$ with width $\phi_1 - \phi_0$. Note that $C(t; \phi) := C(t; 0, \phi)$, $0 < \phi \leq \pi$, is the spherical cap with center t and radius ϕ. Finally we define the function $u(x) := \sqrt{x} J_\alpha(x)$, $x \in \mathbb{R}$, for $\alpha \geq 0$ and assume that $0 = j_{\alpha,0} < j_{\alpha,1} < j_{\alpha,2} < \dots$ are its nonnegative zeros. They satisfy

$$2j_{\alpha,1} > j_{\alpha,2} - j_{\alpha,1} \tag{7}$$

for $\alpha \in \{\frac{n}{2} \mid n \in \mathbb{N}_0\}$. This is well known for $\alpha \in \{0, \frac{1}{2}\}$, and for $\alpha > \frac{1}{2}$ it follows from $j_{\alpha,k+1} - 2j_{\alpha,k} + j_{\alpha,k-1} < 0$, $k \in \mathbb{N}$, which can be proved by applying Sturm's method to the differential equations of $u(x)$ and $v(x) := u(x + j_{\alpha,k-1} - j_{\alpha,k})$ with respect to the interval $j_{\alpha,k} < x < 2j_{\alpha,k} - j_{\alpha,k-1}$. Note that

$$\chi_{\mu k} \sim \frac{1}{\mu} \cdot j_{\alpha k}, \quad \bar{\chi}_{\mu k} \sim \frac{1}{\mu} \cdot j_{\alpha+1,k} \tag{8}$$

holds for fixed $k \in \mathbb{N}$, $\alpha := \frac{r-3}{2}$ and $\mu \to \infty$.

In what follows we always consider a quadrature rule (2) with positive weights and degree of exactness defined by (3).

Example 1 (Weighted Nodes in a Cap): For some constant (which depends on r only) the following inequality holds for $\mu \geq 2$, $\nu := \lfloor \frac{\mu}{2} \rfloor$, and $x \in S^{r-1}$:

$$\sum_{t_j \in C(x; \bar{\chi}_{\nu 1})} A_j \leq \text{const} \cdot \frac{1}{\nu^{r-1}}. \tag{9}$$

We proved this in [2]. The right side has the order of $meas(C(x; \bar{\chi}_{\nu 1}))$, see (8).

Example 2 (Nodes in a Cap): Let $\mu \geq 2$, $\nu := \lfloor \frac{\mu}{2} \rfloor$. Then

$$S^{r-1} \subset \bigcup_{j=1}^{M} C(t_j; \chi_{\nu,1}). \tag{10}$$

This covering property was proved first for spherical designs (equal weights) by Vladimir Yudin [4], but holds also in case of arbitrary positive weights, as pointed out by the author [1]. It can be reformulated in the form:

$$\text{for all } x \in S^{r-1}: \quad C(x; \chi_{\nu 1}) \cap T \neq \emptyset. \tag{11}$$

The assertion is trivial if x is a node. This is different in

Example 3 (Distance of Nodes): Let $\mu \geq 2$, $\nu := \lfloor \frac{\mu}{2} \rfloor$. Then

$$\text{for all } t \in T : \ (C(t; \bar{\chi}_{\nu 1}) \setminus \{t\}) \cap T \neq \emptyset. \tag{12}$$

Proof: First let $x \in S^{r-1}$. By the reproducing property of P_ν we get

$$\int_{S^{r-1}} P_\nu(tx) \left[\frac{P_\nu(tx)}{tx - \bar{\eta}_{\nu 1}} (tx - 1) \right] d\omega(t) = \left[\frac{P_\nu(tx)}{tx - \bar{\eta}_{\nu 1}} (tx - 1) \right]_{t:=x} = 0,$$

where we used that [ ] is a polynomial of degree ν with respect to t. Evaluating the left side by $\hat{I}$ we get in view of (3)

$$\sum_{k=1}^{M} A_k \left[\frac{P_\nu^2(tx)}{tx - \bar{\eta}_{\nu 1}} \right]_{t:=t_k} (t_k x - 1) = 0.$$

Next let $x = t_j \in T$. And assume that $t_j t_k < \bar{\eta}_{\nu,1}$ holds for all $k \neq j$. Then

$$\sum_{k \neq j} A_k \cdot P_\nu^2(t_j t_k) \cdot \frac{t_j t_k - 1}{t_j t_k - \bar{\eta}_{\nu 1}} = 0.$$

must hold where the first and the last factor of the sum members are positive. This implies $P_\nu^2(t_j t_k) = 0$ for $k \neq j$ and hence the contradiction

$$0 < \int_{S^{r-1}} \left[\frac{P_\nu(t_j t)}{t_j t - \bar{\eta}_{\nu 1}} \right]^2 (1 - t_j t) \, d\omega(t) = \sum_{k-1}^{M} A_k \left[\frac{P_\nu(t_j t)}{t_j t - \bar{\eta}_{\nu 1}} \right]^2_{t:=t_k} (1 - t_j t_k) = 0.$$

So the assumption was false, there is some $k \neq j$ such that $t_j t_k \geq \bar{\eta}_{\nu 1} = \cos \bar{\chi}_{\nu 1}$ holds, and as t_j has been arbitrary, (12) is valid.

Examples 4 and 5 (Nodes in a Collar): Applying our method to

$$\int_{S^{r-1}} \frac{G_{\nu+1}^2(tx) \, d\omega(t)}{(tx - \eta_{\nu+1,k+1})(tx - \eta_{\nu+1,k})} = 0 \quad \text{and} \quad \int_{S^{r-1}} \frac{P_\nu^2(tx) \, d\omega(t)}{(tx - \bar{\eta}_{\nu,k+1})(tx - \bar{\eta}_{\nu,k})} = 0,$$

respectively, we similarly obtain the following results:

Example 4: for all $x \in S^{r-1} : \ C(x; \chi_{\nu+1,k}, \chi_{\nu+1,k+1}) \cap T \neq \emptyset$, (13)

Example 5: for all $x \in S^{r-1} : \ C(x; \bar{\chi}_{\nu+1,k}, \bar{\chi}_{\nu+1,k+1}) \cap T \neq \emptyset$. (14)

We remark the following: From (8) and (7) we get $2\chi_{\nu 1} > \chi_{\nu 2} - \chi_{\nu 1}$ for large ν, saying that no collar (13) contains a cap $C(x; \chi_{\nu 1})$. Therefore Example 4 is not covered by Example 2. Nor is Example 5 covered by Example 2. This follows likewise for $k = 1$ and $r = 3$ or $r = 5$ from the inequalities

$$2j_{0,1} > j_{1,2} - j_{1,1} \quad and \quad 2j_{1,1} > j_{2,2} - j_{2,1},$$

which hold by results of Watson [3], p.748.

In the case $r = 3$ we find that every $x \in S^2$ has a neighbour t_k in the distance $\chi_{\nu 1} \sim \frac{1}{\nu} \cdot j_{0,1}$, while t_k has a neighbour $t_l \neq t_k$ in distance $\bar{\chi}_{\nu 1} \sim \frac{1}{\nu} \cdot j_{1,1}$. But

$$\lim_{\nu \to \infty} \frac{\bar{\chi}_\nu}{\chi_\nu} = \frac{j_{1,1}}{j_{0,1}} = 1.59....$$

So $\frac{\bar{\chi}_\nu}{\chi_\nu}$ approximates the golden ratio $1.61... = \frac{1}{2}(1 + \sqrt{5})$, which is the ratio of a diagonal and the radius in a regular pentagon. By chance only? Or is this a hint that the confluent squares, pentagons and hexagons, which can be observed in actual node configurations, could be explained by approximately regular pentagons, at least under additional assumptions, such as minimal node number or maximal determinant?

References

[1]   M. Reimer: *Spherical polynomial approximations: A survey.* In: Advances in Multivariate Approximation, W. Haußmann, K. Jetter, M. Reimer (eds.), Wiley–VCH, Berlin 1999, pp. 231–252.

[2]   M. Reimer: *Hyperinterpolation on the sphere at the minimal projection order*, J. Approx. Theory **104** (2000), 272–286.

[3]   G. N. Watson: *A treatise on the theory of Bessel functions, 2nd ed.* Cambridge Univ. Press, Cambridge 1966.

[4]   V. Yudin: *Covering a sphere and extremal properties of orthogonal polynomials*, Discrete Math. Appl. **5** (1995), 371–379.

Address:

Manfred Reimer
Fachbereich Mathematik
Universität Dortmund
D–44221 Dortmund
Germany

International Series of Numerical Mathematics
Vol. 137, ©2001 Birkhäuser Verlag Basel/Switzerland

Normalized Tight Frames
in Finite Dimensions

Georg Zimmermann

Abstract

For any dimension d and for any $n \geq d$, we construct normalized tight frames with n elements for $\mathbb{C}^d$ and for $\mathbb{R}^d$.

The Problem

In one of the problem sessions, the following question was raised.

Consider frames for $\mathbb{R}^3$ with $n \geq 3$ elements and the property that every frame vector has Euclidean norm 1. What is the minimal quotient of the frame bounds for each n?

We answer this question in full generality by showing that for any $\mathbb{R}^d$ and any $n \geq d$, there exist tight frames with the desired property, so for any d, the minimal quotient of the frame bounds equals 1 for each $n \geq d$.

1 Definitions and Preliminaries

Definition. A *frame* for a Hilbert space $\mathbf{H}$ is a family of vectors $\{v_k\}_{k \in I}$ in $\mathbf{H}$ with the property

$$C_1 \|x\|_{\mathbf{H}}^2 \leq \sum_{k \in I} |\langle x, v_k \rangle|^2 \leq C_2 \|x\|_{\mathbf{H}}^2 \qquad \text{for all } x \in \mathbf{H},$$

where the *frame bounds* C_1, C_2 satisfy $0 < C_1 \leq C_2 < \infty$.
The frame is *tight*, if $C_1 = C_2 =: C$. We say that the frame is *normalized*, if its elements satisfy $\|v_k\|_{\mathbf{H}} = 1$ for all $k \in I$.
If $\mathbf{H} = \mathbb{K}^d$ where $\mathbb{K} = \mathbb{R}$ or $\mathbb{C}$, we define the *matrix* of a finite family with n elements in $\mathbf{H}$ to be the $d \times n$-matrix whose columns are the coefficient vectors

of the elements with respect to the canonical basis.

As usual, we write $A^* = \overline{A}^T$ for the conjugate transpose of a matrix.

The following results can be considered folklore.

Lemma 1.1. (i) *A family* $\{v_k\}_{k=1\ldots n}$ *in* $\mathbb{K}^d$ *is a tight frame with frame bound* C *if and only if its matrix* A *satisfies*

$$A \cdot A^* = C\, I_d\,.$$

(ii) *The frame is normalized if and only if the diagonal of* $A^* \cdot A$ *equals* $(1, \ldots, 1)$.

Proof. (i) Note that for $x = (x_1, \ldots, x_d)^T \in \mathbb{K}^d$, we have

$$A^* x = (\langle x, v_1 \rangle, \ldots, \langle x, v_n \rangle)^T\,.$$

Therefore, the family $\{v_k\}_{k=1\ldots n}$ is a tight frame if and only if

$$C\, \|x\|_{\mathbf{H}}^2 = \sum_{k=1}^{n} |\langle x, v_k \rangle|^2 = x^* \cdot A \cdot A^* \cdot x \qquad \text{for all } x \in \mathbb{K}^d.$$

This shows the claim.

(ii) Obvious. $\square$

Corollary 1.2. *A normalized tight frame of* $\mathbb{K}^d$ *with* n *elements has frame bound* $C = \frac{n}{d}$.

Proof. Denoting the trace of a matrix M by $\mathrm{tr}(M)$, we have for the frame bound C of a normalized tight frame with frame matrix A the equality

$$C = \tfrac{1}{d} \mathrm{tr}(A \cdot A^*) = \tfrac{1}{d} \mathrm{tr}(A^* \cdot A) = \tfrac{1}{d}\, n\,.$$

$\square$

2 Normalized Tight Frames for $\mathbb{C}^d$

Lemma 1.1 shows immediately how to construct normalized tight frames with n elements for $\mathbb{C}^d$. Consider the Fourier matrix

$$F_n = \left(e^{2\pi i j k / n} \right)_{j,k=1\ldots n}$$

which satisfies $F_n \cdot F_n^* = n\, I_n$. Any d rows of F_n form a $d \times n$-matrix B with $B \cdot B^* = n\, I_d$, where every column has norm $\sqrt{d}$, so $A = \frac{1}{\sqrt{d}} B$ is the matrix of a normalized tight frame for $\mathbb{C}^d$.

3 Normalized Tight Frames for $\mathbb{R}^d$

In the real case, we make use of a real–valued version of F_n.

For odd values $n = 2k+1$, define

$$
CS_n =
\begin{pmatrix}
1/\sqrt{2} & 1/\sqrt{2} & 1/\sqrt{2} & \cdots & 1/\sqrt{2} \\
1 & \cos(2\pi\frac{1}{n}) & \cos(2\pi\frac{2}{n}) & \cdots & \cos(2\pi\frac{n-1}{n}) \\
0 & \sin(2\pi\frac{1}{n}) & \sin(2\pi\frac{2}{n}) & \cdots & \sin(2\pi\frac{n-1}{n}) \\
1 & \cos(2\pi\frac{2}{n}) & \cos(2\pi\frac{4}{n}) & \cdots & \cos(2\pi\frac{2(n-1)}{n}) \\
0 & \sin(2\pi\frac{2}{n}) & \sin(2\pi\frac{4}{n}) & \cdots & \sin(2\pi\frac{2(n-1)}{n}) \\
\vdots & \vdots & \vdots & & \vdots \\
1 & \cos(2\pi\frac{k}{n}) & \cos(2\pi\frac{k2}{n}) & \cdots & \cos(2\pi\frac{k(n-1)}{n}) \\
0 & \sin(2\pi\frac{k}{n}) & \sin(2\pi\frac{k2}{n}) & \cdots & \sin(2\pi\frac{k(n-1)}{n})
\end{pmatrix},
$$

while for $n = 2k$, let

$$
CS_n =
\begin{pmatrix}
1/\sqrt{2} & 1/\sqrt{2} & 1/\sqrt{2} & \cdots & 1/\sqrt{2} \\
1 & \cos(2\pi\frac{1}{n}) & \cos(2\pi\frac{2}{n}) & \cdots & \cos(2\pi\frac{n-1}{n}) \\
0 & \sin(2\pi\frac{1}{n}) & \sin(2\pi\frac{2}{n}) & \cdots & \sin(2\pi\frac{n-1}{n}) \\
1 & \cos(2\pi\frac{2}{n}) & \cos(2\pi\frac{4}{n}) & \cdots & \cos(2\pi\frac{2(n-1)}{n}) \\
0 & \sin(2\pi\frac{2}{n}) & \sin(2\pi\frac{4}{n}) & \cdots & \sin(2\pi\frac{2(n-1)}{n}) \\
\vdots & \vdots & \vdots & & \vdots \\
1 & \cos(2\pi\frac{k-1}{n}) & \cos(2\pi\frac{(k-1)2}{n}) & \cdots & \cos(2\pi\frac{(k-1)(n-1)}{n}) \\
0 & \sin(2\pi\frac{k-1}{n}) & \sin(2\pi\frac{(k-1)2}{n}) & \cdots & \sin(2\pi\frac{(k-1)(n-1)}{n}) \\
+1/\sqrt{2} & -1/\sqrt{2} & +1/\sqrt{2} & \cdots & -1/\sqrt{2}
\end{pmatrix}.
$$

These matrices satisfy $CS_n \cdot CS_n^* = \frac{n}{2} I_n$. To generate a normalized tight frame for $\mathbb{R}^d$ with $n \geq d$ elements, start from CS_n. If $d = n$, let $B = CS_n$. If $d < n$, choose d rows in the following manner: For even d, take $\frac{d}{2}$ pairs of corresponding cosine- and sine-rows; if d is odd, in addition include the top row. These rows form a $d \times n$-matrix B satisfying $B \cdot B^* = \frac{n}{2} I_d$. Furthermore, for even d, each column has norm

$$\sqrt{\sum_{j=1}^{d/2}(\cos^2\alpha_j+\sin^2\alpha_j)} = \sqrt{\frac{d}{2}}\,;$$

if d is odd, we obtain

$$\sqrt{\frac{1}{2} + \sum_{j=1}^{(d-1)/2}(\cos^2\alpha_j+\sin^2\alpha_j)} = \sqrt{\frac{d}{2}}\,.$$

Consequently, in each case, $A = \sqrt{\frac{2}{d}}\,B$ is the matrix of a tight normalized frame.

Acknowledgments. The author would like to thank Hans G. Feichtinger for suggesting the problem and Götz Pfander for stimulating discussions.

Address:

Georg Zimmermann
Institut für Angewandte Mathematik und Statistik
Universität Hohenheim
D–70593 Stuttgart
Germany

International Series of Numerical Mathematics
Vol. 137, ©2001 Birkhäuser Verlag Basel/Switzerland

Publications of Jochen W. Schmidt [1]

[1]   J. W. Schmidt: *Fehlerabschätzungen für eine Klasse von Iterationsverfahren.* Z. Angew. Math. Mech. **39** (1959), 392–394.

[2]   J. W. Schmidt: *Konvergenzuntersuchungen und Fehlerabschätzungen für ein verallgemeinertes Iterationsverfahren.* Arch. Rational Mech. Anal. **6** (1961), 261–276.

[3]   J. W. Schmidt: *Regula Falsi für Operatoren in Banachräumen.* Z. Angew. Math. Mech. **41** (1961), T61–T63.

[4]   J. W. Schmidt: *Konvergenzaussagen und Fehlerabschätzungen für ein verallgemeinertes Iterationsverfahren.* In: Proceed. II. Ungar. Math. Kongreß, Vol. II, 1961, 36–39.

[5]   J. W. Schmidt: *Fehlerabschätzungen mit Hilfe eines Vergleichssatzes.* Z. Angew. Math. Mech. **42** (1962), 187–193.

[6]   J. W. Schmidt, H. Schönheinz: *Fehlerschranken zum Differenzenverfahren unter ausschließlicher Benutzung verfügbarer Größen.* Arch. Rational Mech. Anal. **10** (162), 311–322.

[7]   J. W. Schmidt: *Eine Übertragung der Regula Falsi auf Gleichungen in Banachräumen I.* Z. Angew. Math. Mech. **43** (1963), 1–8.

[8]   J. W. Schmidt: *Eine Übertragung der Regula Falsi auf Gleichungen in Banachräumen II.* Z. Angew. Math. Mech. **43** (1963), 97–110.

[9]   J. W. Schmidt: *Ein Vergleichssatz unter Verwendung höherer Ableitungen.* Z. Angew. Math. Mech. **43** (1963), 81–83.

[10]   J. W. Schmidt: *Zur Fehlerabschätzung näherungsweiser Lösungen von Gleichungen in halbgeordneten Räumen.* Arch. Math. **14** (1963), 130–138.

[11]   J. W. Schmidt: *Extremwertermittlung mit Funktionswerten.* Wiss. Z. Tech. Univ. Dresden **12** (1963), 1601–1605.

[12]   J. W. Schmidt: *Ausgangsvektoren für monotone Iterationen bei linearen Gleichungssystemen.* Numer. Math. **6** (1964), 78–88.

[1]Compiled by Marion Bastian, Bernd Mulansky, Gisela Terno, Torsten Schütze and Hubert Schwetlick, Technical University of Dresden, Institute for Numerical Mathematics

[13]   J. W. Schmidt: *Konvergenzbeschleunigung bei monotonen Vektorfolgen.*
         Acta Math. Acad. Sci. Hungar. **16** (1965), 221–229.

[14]   J. W. Schmidt: *Fehlerabschätzung und Konvergenzbeschleunigung zu Ite-
         rationen bei linearen Gleichungssystemen.* Apl. Mat. **19** (1965), 297–
         301.

[15]   J. W. Schmidt: *Asymptotische Einschließung bei konvergenzbeschleu-
         nigenden Verfahren.* Numer. Math. **8** (1966), 105–113.

[16]   J. W. Schmidt: *Konvergenzgeschwindigkeit der Regula Falsi und des
         Steffensen-Verfahrens im Banachraum.* Z. Angew. Math. Mech. **46**
         (1966), 146–148.

[17]   J. W. Schmidt: *Defektabschätzungen bei Differenzenverfahren.* Z.
         Angew. Math. Mech. **46** (1966), 17–39.

[18]   J. W. Schmidt, H.-F. Trinkaus: *Extremwertermittlung mit Funktions-
         werten bei Funktionen von mehreren Veränderlichen.* Computing **1**
         (1966), 224–232.

[19]   J. W. Schmidt, H. Dreßel: *Fehlerabschätzungen bei Polynomgleichungen
         mit dem Fixpunktsatz von Brouwer.* Numer. Math. **10** (1967), 42–50.

[20]   J. W. Schmidt: *Asymptotische Einschließung bei konvergenzbeschleuni-
         genden Verfahren II.* Numer. Math. **11** (1968), 53–56.

[21]   J. W. Schmidt: *Ein Konvergenzsatz für Iterationsverfahren.* Math.
         Nachr. **37** (1968), 67–82.

[22]   J. W. Schmidt, H. Schönheinz: *Fehlerschranken für die genäherte
         Lösung von Rand- und Eigenwertaufgaben bei gewöhnlichen Differ-
         entialgleichungen durch Differenzenverfahren.* In: Numerische Math-
         ematik, Differentialgleichungen, Approximationstheorie, L. Collatz,
         G. Meinardus, H. Unger (eds.), Internat. Ser. Numer. Math. **9**,
         Birkhäuser, Basel 1968, pp. 125–140.

[23]   J. W. Schmidt, H. Schwetlick: *Ableitungsfreie Verfahren mit höherer
         Konvergenzgeschwindigkeit.* Computing **3** (1968), 215–226.

[24]   J. W. Schmidt: *Monotone Einschließung mit der Regula Falsi bei kon-
         vexen Funktionen.* Z. Angew. Math. Mech. **50** (1970), 640–643.

[25]   J. W. Schmidt, D. Leder: *Ableitungsfreie Verfahren ohne Auflösung
         linearer Gleichungen.* Computing **5** (1970), 71–81.

[26]   J. W. Schmidt, H. Leonhardt: *Eingrenzung von Lösungen mit Hilfe der
         Regula Falsi.* Computing **6** (1970), 318–328.

[27]   J. W. Schmidt, K. Vetters: *Ableitungsfreie Verfahren für nichtlineare
         Optimierungsprobleme.* Numer. Math. **15** (1970), 263–282.

[28]   J. W. Schmidt: *Eingrenzung von Lösungen nichtlinearer Gleichungen durch Verfahren mit höherer Konvergenzgeschwindigkeit.* Computing **8** (1971), 208–215.

[29]   J. W. Schmidt: *Eindeutige Lösbarkeit von linearen Optimierungsproblemen.* Z. Angew. Math. Mech. **51** (1971), 153–155.

[30]   J. W. Schmidt: *Einschließung von Nullstellen bei Operatoren mit monoton zerlegbarer Steigung durch überlinear konvergente Iterationsverfahren.* Annal. Acad. Sci. Fennicae, AI **502** (1972), 1–15.

[31]   J. W. Schmidt: *Überlinear konvergente Mehrschrittverfahren vom Regula Falsi- und Newton-Typ.* Z. Angew. Math. Mech. **53** (1973), 103–114.

[32]   F. Kuhnert, J. W. Schmidt: *Numerische Mathematik.* In: Entwicklung der Mathematik in der DDR, Deutscher Verlag der Wissenschaften, Berlin 1974, pp. 347–399.

[33]   J. W. Schmidt: *Regula Falsi-Verfahren mit konsistenter Steigung und Majorantenprinzip.* Period. Math. Hungar. **5** (1974), 187–193.

[34]   J. W. Schmidt: *Ein ableitungsfreies Mehrschrittverfahren für Extremwertaufgaben.* Math. Nachr. **59** (1974), 95–104.

[35]   J. W. Schmidt: *Zur Herleitung der Konvergenzordnung beim gewöhnlichen Differenzenverfahren und beim Mehrstellenverfahren.* In: Mathematical Structures – Computational Mathematics – Mathematical Modelling, Vol. 1, B. Sendov (ed.), Publ. House Bulg. Acad. Sci., Sofia, 1975, pp. 425–438.

[36]   J. W. Schmidt: *Bemerkungen zu einem Verfahren von H. J. Stetter.* Beiträge Numer. Math. **4** (1975), 205–213.

[37]   J. W. Schmidt: *Monotone Einschließung von Inversen positiver Elemente durch Verfahren vom Schulz-Typ.* Computing **16** (1976), 211–219.

[38]   J. W. Schmidt: *Über lineare Ungleichungen vom Gronwallschen Typ.* Beiträge Numer. Math. **5** (1976), 171–189.

[39]   J. W. Schmidt, W. Nauber: *Über Verfahren zur zweiseitigen Approximation inverser Elemente.* Computing **17** (1976), 59–67.

[40]   J. W. Schmidt: *Monotone Einschließung von Inversen positiv zerlegbarer Elemente durch ein quadratisch konvergentes Verfahren.* Abh. Akad. Wiss. DDR, **N1** (1977), 373–379.

[41]   J. W. Schmidt: *Eine Anwendung des Brouwerschen Fixpunktsatzes zur Gewinnung von Fehlerschranken für Näherungen von Polynomnullstellen.* Beiträge Numer. Math. **6** (1977), 158–163.

[42]   J. W. Schmidt: *Ein kubisch konvergentes Einschließungsverfahren zur Inversion positiver Elemente.* Computing **19** (1977), 175–178.

[43]   J. W. Schmidt: *Finitisierung und Globalisierung bei Iterationsverfahren zur Lösung nichtlinearer Gleichungssysteme.* In: Berichte VII. IKM Weimar, Verl. f. Bauwesen, Berlin 1977, pp. 307–317.

[44]   J. W. Schmidt: *Herleitung der Konvergenzordnung bei Differenzenverfahren mit Hilfe Greenscher Funktionen.* Godisnik Viss. Ucebn. Zaved. Prilozna Mat. **11** (1977), 181–190.

[45]   J. W. Schmidt, W. Hoyer: *Ein Konvergenzsatz für Verfahren vom Brown–Brent–Typ.* Z. Angew. Math. Mech. **57** (1977), 397–405.

[46]   J. W. Schmidt: *Untere Fehlerschranken für Regula Falsi–Verfahren.* Period. Math. Hungar. **9** (1978), 241–247.

[47]   J. W. Schmidt: *Zur Konvergenz von kubischen Interpolationssplines.* Z. Angew. Math. Mech. **58** (1978), 109–110.

[48]   J. W. Schmidt: *Selected contributions to imbedding methods for finite dimensional problems.* In: Continuation Methods, H. Wacker (ed.), Academic Press, New York 1978, pp. 215–247.

[49]   J. W. Schmidt: *Einschließung inverser Elemente durch Fixpunktverfahren.* Numer. Math. **31** (1978), 313–320.

[50]   J. W. Schmidt: *Iterative Verbesserung von genäherten LU–Faktorisierungen.* Preprint 07–33–78, Techn. Univ. Dresden, Sektion Mathematik, 1978.

[51]   J. W. Schmidt, W. Hoyer: *Die Verfahren vom Brown–Brent–Typ bei gemischt linearen–nichtlinearen Gleichungssystemen.* Z. Angew. Math. Mech. **58** (1978), 425–428.

[52]   J. W. Schmidt, H. Mettke: *Konvergenz von quadratischen Interpolations- und Flächenabgleichssplines.* Computing **19** (1978), 351–363.

[53]   J. W. Schmidt: *Zur Numerik nichtlinearer Probleme – Aufgaben, Algorithmen und Programme.* Wiss. Z. Tech. Univ. Dresden **28** (1979), 1081–1087.

[54]   J. W. Schmidt, U. Patzke: *Nachiteration und monotone Eingrenzung bei der Cholesky-Faktorisierung.* Preprint 07–30–1979, Techn. Univ. Dresden, Sektion Mathematik, 1979.

[55]   J. W. Schmidt: *Ergebnisse zum Brown–Brent–Verfahren zur numerischen Lösung nichtlinearer Gleichungssysteme.* Wiss. Z. Tech. Univ. Dresden **29** (1980), 422–445.

[56] J. W. Schmidt: *Monotone Eingrenzung von inversen Elementen durch ein quadratisch konvergentes Verfahren ohne Durchschnittsbildung.* Z. Angew. Math. Mech. **60** (1980), 202–204.

[57] J. W. Schmidt: *On the R-order of coupled sequences.* Computing **26** (1981), 33–42.

[58] J. W. Schmidt, U. Patzke: *Iterative Nachverbesserung mit Fehlereingrenzung der Cholesky-Faktoren von Stieltjes-Matrizen.* J. Reine Angew. Math. **327** (1981), 81–92.

[59] W. Burmeister, J. W. Schmidt: *On the R-order of coupled sequences II.* Computing **29** (1982), 73–81.

[60] J. W. Schmidt: *Ein Einschließungsverfahren für Lösungen fehlerbehafteter linearer Gleichungen.* Period. Math. Hungar. **13** (1982), 29–37.

[61] J. W. Schmidt: *Monoton einschließende Näherungsverfahren bei nichtlinearen Gleichungen.* Mitt. Math. Ges. DDR, Heft 1/2, 1982, 62–87.

[62] W. Burmeister, J. W. Schmidt: *On the R-order of coupled sequences III.* Computing **30** (1983), 157–169.

[63] F.-A. Potra, J. W. Schmidt: *On a class of iterative procedures with monotonous convergence.* Numer. Funct. Anal. Optim. **6** (1983), 1–23.

[64] J. W. Schmidt, H. Schneider: *Monoton einschließende Verfahren bei additiv zerlegbaren Gleichungen.* Z. Angew. Math. Mech. **63** (1983), 3–11.

[65] W. Burmeister, J. W. Schmidt: *Über Kegel in endlichdimensionalen Räumen.* Beiträge Numer. Math. **12** (1984), 29–32.

[66] W. Burmeister, J. W. Schmidt: *Determination of the cone radius for positive concave operators.* Computing **33** (1984), 37–49.

[67] W. Hoyer, J. W. Schmidt: *Newton-type decomposition methods for equations arising in network analysis.* Z. Angew. Math. Mech. **64** (1984), 397–405.

[68] J. W. Schmidt: *Two-sided approximations of inverses, square roots and Cholesky factors.* In: Computational Mathematics of Banach Center Publ. **13**, A. Wakulicz (ed.), Polish Scient. Publishers, Warsaw 1984, pp. 483–497.

[69] J. W. Schmidt, W. Heß: *Schwach verkoppelte Ungleichungssysteme und konvexe Spline-Interpolation.* Elem. Math. **39** (1984), 85–95.

[70] J. W. Schmidt, H. Schneider: *Enclosing methods in perturbed nonlinear operator equations.* Computing **32** (1984), 1–11.

[71]   W. Burmeister, W. Heß, J. W. Schmidt: *Convex spline interpolants with minimal curvature.* Computing **35** (1985), 219–229.

[72]   W. Burmeister, J. W. Schmidt: *Characterization of the best R–orders of coupled sequences arising in iterative processes.* In: Proc. Internat. Conf. Numer. Math. and Appl. **84**, Publ. House Bulg. Acad. Sci., Sofia 1985, 191–202.

[73]   W. Burmeister, J. W. Schmidt: *Ermittlung der R–Ordnung von iterativen Näherungsverfahren.* Mitt. Math. Ges. DDR, Heft 2/3, 1985, 55–58.

[74]   E. Neuman, J. W. Schmidt: *On the convergence of quadratic spline interpolants.* J. Approx. Theory **45** (1985), 299–309.

[75]   J. W. Schmidt: *Eine Aufgabe von J. W. Schmidt.* Schülerzeitschrift alpha, 1985, p. 85.

[76]   J. W. Schmidt, W. Hoyer, C. Haufe: *Consistent approximations in Newton–type decomposition algorithms.* Numer. Math. **47** (1985), 413–425.

[77]   W. Heß, J. W. Schmidt: *Convexity preserving interpolation with exponential splines.* Computing **36** (1986), 335–342.

[78]   J. W. Schmidt: *Convex interval interpolation with cubic splines.* BIT **26** (1986), 377–387.

[79]   J. W. Schmidt: *Bestimmung minimal gekrümmter Splines mit Hilfe dualer Aufgaben.* Seminarbericht **80**, Humboldt–Univ. Berlin, Sektion Mathematik, 1986.

[80]   J. W. Schmidt: *Enclosing methods for perturbed boundary value problems in nonlinear difference equations.* In: Proceedings EQUADIFF 6, J. Vosmansky, M. Zlamal (eds.), Brno 1995, Lecture Notes Math. **1192**, Springer, Berlin 1986, pp. 333–338.

[81]   W. Burmeister, S. Dietze, W. Heß, J. W. Schmidt: *Solution of a class of weakly coupled programs via dualization, and applications.* In: Discretization in Differential Equations and Enclosures of Math. Research **36**, E. Adams, R. Ansorge, Ch. Großmann, H.–G. Roos (eds.), Akademie Verlag, Berlin 1987, pp. 57–80.

[82]   J. W. Schmidt: *An unconstrained dual program for computing convex C^1–spline approximants.* Computing **39** (1987), 133–140.

[83]   J. W. Schmidt: *On convex cubic C^2–spline interpolation.* In: Numerical Methods of Approximation Theory, L. Collatz, G. Meinardus, G. Nürnberger (eds.), Internat. Ser. Numer. Math. **81**, Birkhäuser, Basel 1987, pp. 213–228.

[84]   J. W. Schmidt: *On tridiagonal linear complementarity problems.* Numer. Math. **51** (1987), 11–21.

[85]   J. W. Schmidt: *Specially structured convex optimization problems: Computational aspects and applications.* In: Numerical Methods, Colloquia Math. Soc. Janos Bolyai **50**, D. Greenspan, P. Rózsa (eds.), North Holland 1987, pp. 565–579.

[86]   J. W. Schmidt: *A class of superlinear decomposition methods in nonlinear equations.* Numer. Funct. Anal. Optim. **9** (1987), 629–645.

[87]   J. W. Schmidt, W. Heß: *Quadratic and related exponential splines in shape preserving interpolation.* J. Comput. Appl. Math. **18** (1987), 321–329.

[88]   J. W. Schmidt, W. Heß: *Positive interpolation with rational quadratic splines.* Computing **38** (1987), 261–267.

[89]   W. Burmeister, J. W. Schmidt: *A lemma on convex functionals in finite-dimensional spaces.* Arch. Math. (Basel) **50** (1988), 189–192.

[90]   W. Burmeister, J. W. Schmidt: *On the R–order of coupled sequences arising in single–step type methods.* Numer. Math. **53** (1988), 653–661.

[91]   S. Dietze, J. W. Schmidt: *Determination of shape preserving spline interpolants with minimal curvature via dual programs.* J. Approx. Theory **52** (1988), 43–57.

[92]   J. W. Schmidt: *Unconstrained pendants to discretized constrained variational problems.* In: Proceed. ISNA II, Prag 1987, Teubner–Texte **107**, Teubner, Leipzig 1988, pp. 268–272.

[93]   J. W. Schmidt: *Unconstrained dual program to some discretized variational problems with side conditions.* In: Proceed. NUMDIFF 4, Teubner–Texte **104**, Teubner, Leipzig 1988, pp. 232–236.

[94]   J. W. Schmidt, W. Heß: *Positivity of cubic polynomials on intervals and positive spline interpolation.* BIT **28** (1988), 340–352.

[95]   S. Dietze, J. W. Schmidt: *Unconstrained dual programs for partially separable constrained optimization problems.* Preprint 07–10–89, Techn. Univ. Dresden, Sektion Mathematik, 1989.

[96]   W. Hoyer, J. W. Schmidt, N. Shabani: *Superlinearly convergent decomposition methods for block–tridiagonal nonlinear systems of equations.* Numer. Funct. Anal. Optim. **10** (1989), 961–975.

[97]   M. Sakai, J. W. Schmidt: *Positive interpolation with rational splines.* BIT **29** (1989), 140–147.

[98]   J. W. Schmidt: *On shape preserving spline interpolation: Existence the-orems and determination of optimal splines.* In: Approximation and Function Spaces, Banach Center Publ. **22**, Z. Ciesielski (ed.), Polish Scient. Publishers, Warsaw 1989, pp. 377–389.

[99]   J. W. Schmidt: *Results and problems in shape preserving interpolation and approximation with polynomial splines.* In: Splines in Numerical Analysis, Proceed. ISAM Weißig 89, J. W. Schmidt, H. Späth (eds.), Math. Research **52**, Akademie Verlag, Berlin 1989, pp. 159–170.

[100]   J. W. Schmidt: *Beiträge zur Konvergenzordnung bei Iterationsverfahren.* Nova Acta Leopoldina (N.F.) 61, **267** (1989), 97–105.

[101]   J. W. Schmidt: *Convex smoothing by splines and dualization.* In: Proc. Internat. Conf. Numerical Methods and Applications 88, Publ. House Bulg. Acad. Sci., Sofia 1989, 425–436.

[102]   J. W. Schmidt, W. Heß: *Spline interpolation under two–sided restrictions on the derivatives.* Z. Angew. Math. Mech. **69** (1989), 353–365.

[103]   W. Heß, J. W. Schmidt: *Formerhaltende Histopolation mit Exponential-splines.* Preprint 07–15–1990, Techn. Univ. Dresden, Sektion Mathe-matik, 1990.

[104]   J. W. Schmidt: *Monotone data smoothing by quadratic splines via du-alization.* Z. Angew. Math. Mech. **70** (1990), 299–307.

[105]   J. W. Schmidt: *Numerical solution of some classes of complementarity and variational problems.* In: Numerical Analysis and Mathematical Modelling, Banach Center Publ. **24**, Polish Scient. Publishers, Warsaw 1990, 165–179.

[106]   J. W. Schmidt, W. Heß, T. Nordheim: *Shape preserving histopolation using rational quadratic splines.* Computing **44** (1990), 245–258.

[107]   J. W. Schmidt, M. Sakai: *A criterion for the positivity of rational cubic C^2-spline interpolants.* Computing **44** (1990), 365–368.

[108]   J. W. Schmidt, I. Scholz: *A dual algorithm for convex-concave data smoothing by cubic C^2-splines.* Numer. Math. **57** (1990), 333–350.

[109]   J. W. Schmidt: *Convex interval interpolation with cubic splines II.* BIT **31** (1991), 328–340.

[110]   J. W. Schmidt: *Rational biquadratic C^1-splines in S-convex interpola-tion.* Computing **47** (1991), 87–96.

[111]   J. W. Schmidt: *Beiträge zur konvexen Interpolation, Histopolation und Approximation durch Spline-Funktionen.* Mitt. Math. Ges. Hamburg **12** (1991), 603–628.

[112] J. W. Schmidt, W. Heß: *Lösungsansätze für restringiertes Ausgleichen von Histogrammen durch Spline–Funktionen.* Preprint 07–01–91, Techn. Univ. Dresden, Sektion Mathematik, 1991.

[113] J. W. Schmidt: *Constrained smoothing of histograms by quadratic splines.* Computing **48** (1992), 97–107.

[114] J. W. Schmidt: *Dual algorithms for solving convex partially separable optimization problems.* Jber. Dt. Math.–Verein. **94** (1992), 40–62.

[115] J. W. Schmidt: *Vereinheitlichte Darlegung von dualen Algorithmen zum restringierten Ausgleichen von Punktmengen und Histogrammen durch quadratische Spline–Funktionen.* Wiss. Z. Tech. Univ. Dresden **41** (1992), 68–74.

[116] J. W. Schmidt: *Dual method for smoothing histograms using nonnegative C^1-splines.* In: Numerical Methods of Approximation Theory, D. Braess, L. L. Schumaker (eds.), Internat. Ser. Numer. Math. **105**, Birkhäuser, Basel 1992, pp. 317–329.

[117] J. W. Schmidt: *Positive, monotone, and S-convex C^1-interpolation on rectangular grids.* Computing **48** (1992), 363–371.

[118] J. W. Schmidt, S. Dietze: *Unconstrained duals for partially separable constrained programs.* Math. Programming **56** (1992), 337–341.

[119] B. Mulansky, J. W. Schmidt: *Nonnegative C^1-interpolation of scattered data using Powell–Sabin–splines.* Preprint MATH–NM–02–1993, Techn. Univ. Dresden 1993.

[120] J. W. Schmidt: *Positive, monotone, and S-convex C^1-histopolation on rectangular grids.* Computing **50** (1993), 19–30.

[121] J. W. Schmidt, W. Heß: *S-convex, monotone, and positive interpolation with rational bicubic splines of C^2-continuity.* BIT **33** (1993), 496–511.

[122] J. W. Schmidt, W. Heß: *Shape preserving C^2-spline histopolation.* J. Approx. Theory **75** (1993), 325–345.

[123] W. Heß, J. W. Schmidt: *Direct methods for constructing positive spline interpolants.* In: Wavelets, Images and Surface Fitting, P. J. Laurent, A. Le Méhauté, L. L. Schumaker (eds.), A K Peters, Wellesley 1994, pp. 287–294.

[124] W. Heß, J. W. Schmidt: *Positive quartic, monotone quintic C^2-spline interpolation in one and two dimensions.* J. Comput. Appl. Math. **55** (1994), 51–67.

[125] W. Heß, J. W. Schmidt: *Convex C^3 interpolation with quartic splines on threefold refined grids.* Preprint MATH–NM–12–1994, Techn. Univ. Dresden 1994.

[126] B. Mulansky, J. W. Schmidt: *Nonnegative interpolation by biquadratic splines on refined rectangular grids.* In: Wavelets, Images and Surface Fitting, P. J. Laurent, A. Le Méhauté, L. L. Schumaker (eds.), A K Peters, Wellesley 1994, pp. 379–386.

[127] B. Mulansky, J. W. Schmidt: *Powell–Sabin splines in range restricted interpolation of scattered data.* Computing **53** (1994), 137–154.

[128] J. W. Schmidt: *Dual algorithm for convex approximation of histograms using cubic C^1-splines.* In: Numerical Analysis and Mathematical Modelling, Banach Center Publ. **29**, Polish Scient. Publishers, Warsaw 1994, pp. 35–44.

[129] J. W. Schmidt: *Nonlinear splines in convex and S-convex interpolation.* In: Proceed. Second Colloq. Numer. Anal. Plovdiv, D. Bainov, V. Covachev (eds.), VSP, Zeist 1994, pp. 177–186.

[130] J. W. Schmidt: *Positive and S-convex C^1-interpolation of gridded three dimensional data.* Numer. Funct. Anal. Optim. **16** (1995), 233–246.

[131] J. W. Schmidt: *Staircase algorithm and convex spline interpolation.* Z. Angew. Math. Mech. **75** (1995), S 661–S 662.

[132] J. W. Schmidt: *Computational methods in strip interpolation using spline functions.* Preprint MATH–NM–15–1995, Techn. Univ. Dresden, 1995.

[133] J. W. Schmidt: *S-convexity preserving interpolation of gridded three dimensional data using rational C^1-splines.* In: Proceed. Third Colloq. Numer. Anal. Plovdiv, D. Bainov, V. Covachev (eds.), VSP, Zeist 1995, pp. 161–170.

[134] J. W. Schmidt, W. Heß: *An always successful method in univariate convex C^2 interpolation.* Numer. Math. **71** (1995), 237–252.

[135] M. Herrmann, B. Mulansky, J. W. Schmidt: *Scattered data interpolation subject to piecewise quadratic range restrictions.* J. Comp. Appl. Math. **73** (1996), 209–223.

[136] W. Heß, J. W. Schmidt: *Shape preserving C^3 data interpolation and C^2 histopolation with splines on threefold refined grids.* Z. Angew. Math. Mech. **76** (1996), 487–496.

[137] B. Mulansky, J. W. Schmidt: *Constructive methods in convex C^2 interpolation using quartic splines.* Numer. Algorithms **12** (1996), 111–124.

[138] B. Mulansky, J. W. Schmidt, M. Walther: *Tensor product spline interpolation subject to piecewise bilinear lower and upper bounds.* In: Advanced Course on FAIRSHAPE, J. Hoschek, P. D. Kaklis (eds.), B. G. Teubner, Stuttgart 1996, pp. 201–216.

[139] J. W. Schmidt: *Staircase algorithm and construction of convex spline interpolants up to the continuity C^3.* Comput. Math. Appl. **31** (1996), 67–79.

[140] J. W. Schmidt: *Upper bounds for the second order derivatives in convex spline interpolation.* Investigación Oper. **17** (1996), 149–157.

[141] J. W. Schmidt: *Strip interpolations using splines on refined grids.* In: Proceed. Internat. Workshop on Recent Advances in Applied Mathematics, Kuwait 1996, Kuwait Univ., Kuwait 1996, pp. 463–474.

[142] J. W. Schmidt: *Upper bounds for the second order derivatives in convex spline interpolation.* Investigación Oper. **17** (1996), 149–157.

[143] J. W. Schmidt, M. Walther: *Gridded data interpolation with restrictions on the first order derivatives.* In: Multivariate Approximation and Splines, G. Nürnberger, J. W. Schmidt, G. Walz (eds.), Internat. Ser. Numer. Math. **125** Birkhäuser, Basel 1997, pp. 289–305.

[144] J. W. Schmidt: *Interpolation in a derivative strip.* Computing **58** (1997), 377–389.

[145] J. W. Schmidt, W. Heß: *Numerical methods in strip interpolations applying C^1, C^2, and C^3 splines on refined grids.* Mitt. Math. Ges. Hamburg **15** (1997), 107–135.

[146] J. W. Schmidt, M. Walther: *Tensor product splines on refined grids in S-convex interpolation.* In: Multivariate Approximation: Recent Trends and Results, W. Haußmann, K. Jetter, M. Reimer (eds.), Akademie-Verlag, Berlin 1997, pp. 189–202.

[147] J. W. Schmidt, M. Bastian–Walther: *Algorithm for constructing range restricted histosplines.* Numer. Algorithms **17** (1998), 241–260.

[148] J. W. Schmidt, W. Heß: *Fair upper bounds for the curvature in univariate convex interpolation.* BIT **37** (1998), 948–960.

[149] M. Bastian–Walther, J. W. Schmidt: *Shape preserving interpolation by tensor product splines on refined grids.* In: Creating Fair and Shape-Preserving Curves and Surfaces, H. Nowacki, P. D. Kaklis (eds.), Teubner, Stuttgart 1998, pp. 201–217.

[150] M. Bastian–Walther, J. W. Schmidt: *Gregory's rational cubic splines in interpolation subject to derivative obstacles.* In: M. W. Müller, M. D. Buhmann, D. H. Mache, M. Felten (eds.), New Developments in Approximation Theory, Birkhäuser, Basel 1999, pp. 33–47.

[151]   M. Bastian–Walther, J. W. Schmidt: *Range restricted interpolation using Gregory's rational cubic splines.* J. Comput. Appl. Math. **103** (1999), 221–237.

[152]   J. W. Schmidt: *Range restricted interpolation by cubic C^1-splines on Clough–Tocher splits.* In: Advances in Multivariate Approximation, W. Haußmann, K. Jetter, M. Reimer (eds.), Wiley–VCH, Berlin 1999, pp. 253–267.

[153]   J. W. Schmidt: *Tridiagonally separable programs: Unconstrained duals and applications in convex spline approximation.* Aportaciones Mat. Comun. **24** (1999).

[154]   J. W. Schmidt: *Nonlinear splines in convex interpolation and fair bounds for the curvature.* Z. Angew. Math. Mech. **79** (1999), 877–878.

[155]   B. Mulansky, J. W. Schmidt: *Convex interval interpolation using a three-term staircase algorithm.* Numer. Math. **82** (1999), 313–337.

[156]   J. W. Schmidt: *Scattered data interpolation applying rational quadratic C^1 splines on refined triangulations.* Z. Angew. Math. Mech. **80** (2000), 27–33.

[157]   M. Bastian, J. W. Schmidt: *Nonnegative interpolation with Clough–Tocher splines of cubic precision.* To appear in Z. Angew. Math. Mech.

[158]   B. Mulansky, J. W. Schmidt: *Composition based staircase algorithm and constrained interpolation with boundary conditions.* Numer. Math. **85** (2000), 387–408.

[159]   J. W. Schmidt, M. Bastian, B. Mulansky: *Nonnegative volume matching by cubic C^1 splines on Clough–Tocher splits.* To appear in SIAM J. Numer. Anal.

[160]   J. W. Schmidt: *Staircase algorithm and boundary valued convex interpolation by Gregory's splines.* Preprint MATH–NM–04–2000, Techn. Univ. Dresden, 2000.

[161]   J. W. Schmidt: *Univariate strip interpolations by nonlinear parametric splines.* Computing **65** (2000), 323–337.

[162]   J. W. Schmidt, M. Bastian, B. Mulansky: *Construction of nonnegative volume matching splines.* Preprint MATH–NM–14–2000, Techn. Univ. Dresden, 2000.

Proceedings Volumes

[1] J. W. Schmidt, H. Späth (eds.): *Splines in Numerical Analysis.* Proc. ISAM Weißig 89, Math. Research Vol. **52**, Akademie Verlag, Berlin 1989.

[2] P. Rózsa, J. W. Schmidt, B. A. Szabó (eds.): *Proc. Sixth Internat. Conference on Numerical Methods, Miskolc* 1994. Computers Math. Appl. Vol. **31**, Elsevier Sci. Ltd., Oxford 1996.

[3] G. Nürnberger, J. W. Schmidt, G. Walz (eds.): *Multivariate Approximation and Splines*, Birkhäuser, Basel 1997.

Doctoral Students Supervised by J. W. Schmidt

Hubert Schwetlick (1967), Klaus Vetters (1970), Siegfried Scholz (1970), Wolgang Hoyer (1973), Wolfgang Mönch (1974), Ursula Hans (1975), Wolfgang Burmeister (1976), Walter Nauber (1980), Uwe Patzke (1982), Harald Schneider (1984), Nasser Shabani (1989), Frank Schmidt (1996), Marion Bastian (1999).

Postdocs from J. W. Schmidt's Research Group Who Obtained 'Habilitation'

Hubert Schwetlick (1976), Helmut Kleinmichel (1982), Wolfgang Burmeister (1982), Bernd Mulansky (1999).